Basic Plumbing Illustrated

By the Editors of Sunset Books
and Sunset Magazine

Lane Publishing Co. • Menlo Park, California

Acknowledgments
For their helpful advice and assistance, we wish to extend special thanks to Tom Tyler, Edith Martin, and K. Darryll Martin of K. L. Martin Plumbing Co., David Babb, and Ben Altadonna.

Supervising Editor: Patricia Hart Clifford
Research and Text: Buff Bradley

Design and Illustration: Joe Seney
Cover: Photograph by Norman A. Plate

Editor, Sunset Books: David E. Clark

Ninth Printing June 1980
 Library of Congress No. 75-6261. ISBN 0-376-01463-6. Lithographed in the United States.

Contents

How plumbing works

If your plumbing experience has been limited to turning a faucet on and off, you'll be surprised at the simplicity of the system of pipes behind that faucet. Actually, there are three systems: supply, drainage/waste, and venting. Before you begin any plumbing project—large or small—it's a good idea to be familiar with the way these systems work.

• *The supply system* carries water from a well, a storage tank, or from underground water mains into your house and around to all the fixtures (sinks, showers, toilets) and to such appliances as the washing machine and dishwasher.

• *The drainage/waste system* carries waste water and solid waste out of the house to a sewer or a septic tank.

• *The venting system* carries away sewer gases and maintains atmospheric pressure inside the drainpipes. If atmospheric pressure within pipes were not maintained, water in traps could be siphoned away, allowing deadly sewer gases to enter your home. In addition, improper venting could cause waste stoppage in drain lines.

HOW THE SUPPLY SYSTEM WORKS

If you live in the country, you probably get your water from your own well. If you live in town, your water is delivered to you through underground water mains.

The water main connects with your house supply system near your property line. At that point you'll find either a meter and a valve or just a shutoff valve (sometimes called a "stop cock"). If you live in a cold-winter area, there will be only the stop cock, several feet underground under a cover. If you live where winters don't get extremely cold, you'll find a meter just a little below ground level. The valve is used to turn water service off and on. In an emergency, if your home doesn't have its own shutoff valve, you might try to shut off the water here, though it's difficult to operate the valve without special water company equipment. Try turning the valve clockwise with a wrench.

Water is delivered to your supply system under pressure. Normal pressure is about 50 pounds per square inch (abbreviated "psi"), though it can vary greatly—either higher or lower. (For a discussion of water-pressure troubles, see page 28.) From the main, the water supply travels underground directly to your house, probably in 1-inch or ¾-inch pipe. In cold-winter areas, the water meter is just inside the wall in the basement. At this point you may also find a shutoff valve (not every house has one) that you can use to turn off the water supply to the entire house. In cold-winter climates this valve will be just inside the wall; in

Where to turn off the water

Gate valve, *on the line from the water meter to the house, shuts down the entire house water supply.*

Meter valve *requires a wrench to turn. Use this valve to shut off water if your supply line has no gate valve.*

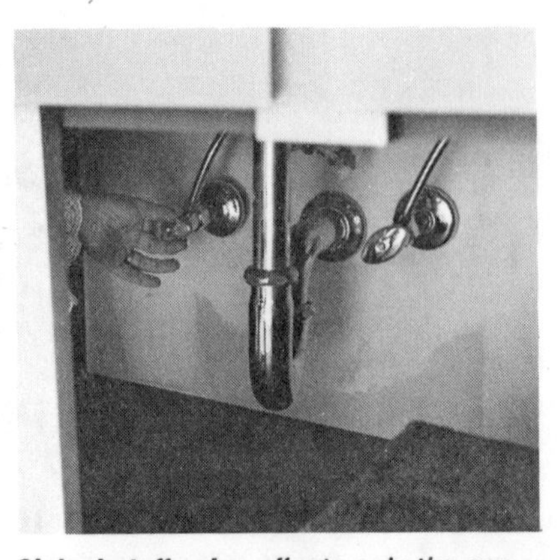

Sink shutoff valve *affects only the connected faucet. A similar valve under the toilet shuts off its water supply.*

Supply system

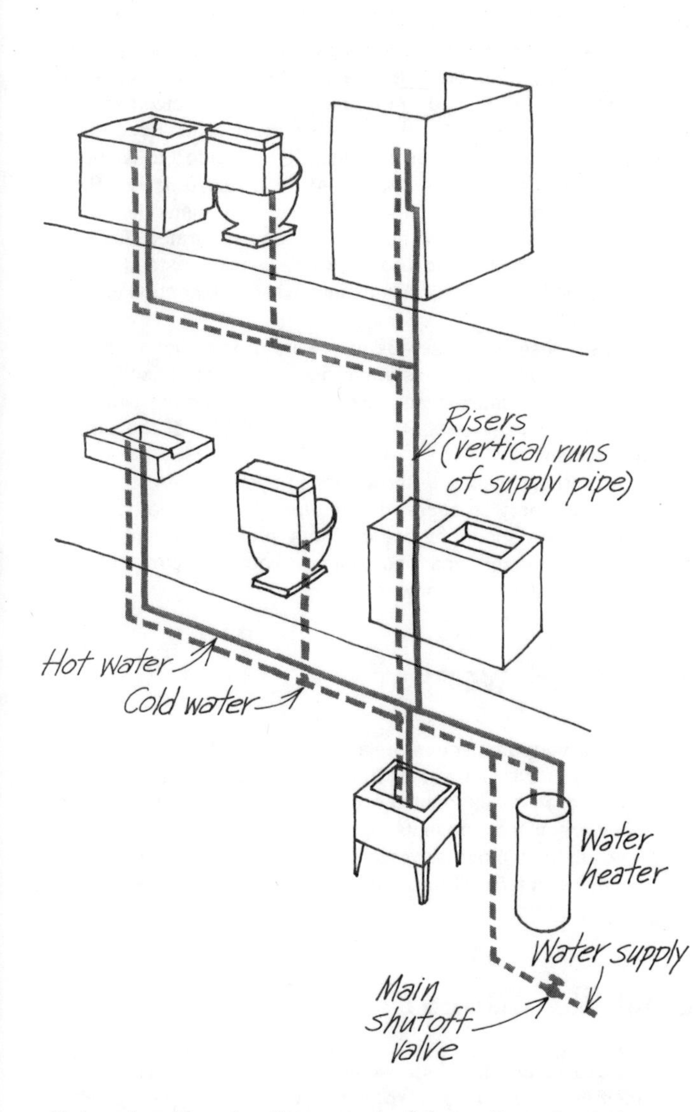

Hot and cold *water lines supply fixtures throughout the house. Look for shutoff valve at each fixture.*

Drain-waste-vent system

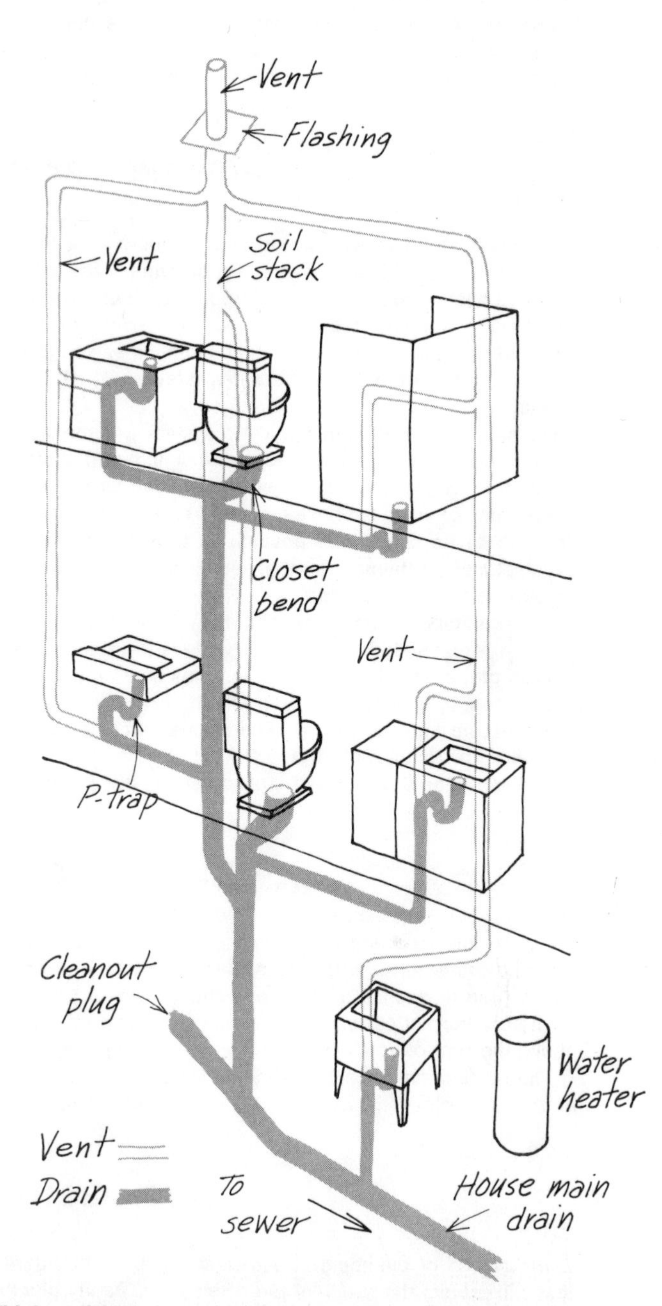

Thickest line *is the main soil stack which drains waste from toilets into the main drain and also serves as a vent.*

moderate climates it may be outside at the wall. It's a good idea for every household member to know its location.

Once inside the house, the pipe branches out into pipes of smaller diameters to deliver water to all outlets. Horizontal pipe runs are generally secured to floor joists, either recessed in notches above the joists or fastened with metal straps below the joists. Long vertical runs can be supported at their bases by platforms and anchored to wall studs.

Supply pipes may be installed with a slight pitch in the runs, sloping back toward the lowest point in the system so that all pipes can be drained. Sometimes at the lowest point there's a valve that can be opened to drain the system.

Some of the water coming into your system goes directly to the cold-water outlets. And some of the water makes an intermediate stop at the water heater, where it is heated and stored.

Most of the fixtures in your home should have individual shutoff valves to enable you to work at one place without having to shut off the supply for the entire house. Look underneath the fixture or behind it for the shutoff valve. You close these valves by turning them clockwise. In an emergency you'll want to go first to the fixture. Everyone living in the house should know how to turn off the water supply, both at individual fixtures and at the main valve.

HOW THE DRAINAGE SYSTEM WORKS

When you're finished with water, you normally want to get rid of it, along with any waste it's carrying. The house drainage system exists for that purpose. (The entire drainage system actually does three jobs—disposing of drain water, carrying away solid wastes, and venting sewer gases. It's commonly called the "drain-waste-vent," or DWV system.)

Of major importance in the DWV system are the traps, the familiar U or S-shaped bends of piping under sinks. Traps are designed to prevent potentially dangerous sewer gases from entering the house; each fixture must have one. Because of its shape and position, a trap, under normal circumstances, retains water that serves as an effective seal against rising sewer gases.

The drainpipes themselves lead away from all fixtures at a carefully calculated slope. If the slope is too steep, water will run off too fast, leaving solid particles behind; if it's not steep enough, water and waste will drain too slowly and may back up into the fixture. The normal pitch is ¼ inch for every horizontal foot of pipe travel.

Central to the DWV system is a soil stack, a vertical section of 3 or 4-inch-diameter pipe that carries waste away from toilets (and often from other fixtures) and connects with the main house drain in the basement or crawl space. The upper part of the stack serves as a vent. Secondary vents from other fixtures can also be connected to it above the level of the highest fixture in the house (this is called reventing or back venting). However, in many homes—especially single-story homes—widely separated fixtures make it impractical to use only one stack. Instead, each fixture or fixture group has its own waste connection and its own vent.

The house drain is a 3-inch, 4-inch, or larger diameter pipe that collects all waste and drainage from the soil stack and other drainpipes and leads out of the house where it joins the house sewer—the underground part of the drainage system.

At various places in the drainage system are cleanouts. Cleanouts provide access to the pipes for clearing obstructions. There should be one cleanout in each horizontal section of drainage line, including an outdoor cleanout for access to the house sewer.

HOW THE VENTING SYSTEM WORKS

Sewer gases can build up enough pressure in your drainpipes to break through the water seal in a trap and enter your house through a drain. The venting system prevents such a buildup of sewer gases by allowing them to escape above the roof of your house.

The venting system also maintains atmospheric pressure in the drainpipes. Water rushing out through traps can create vacuums, causing a siphoning effect that would empty the traps of water. But the constant presence of circulating air throughout the drainage system maintains an equilibrium that prevents siphoning.

All fixtures must be vented. Each may have its own vent or may connect with a main vent through a system of secondary vents.

WHAT ABOUT PLUMBING CODES?

The purpose of plumbing codes is to establish standards that will protect the health of the community. Faulty plumbing is a serious health hazard.

Codes have always varied widely from place to place. Today, with increasing standardization of equipment and materials, plumbing codes in different areas of the country are tending more and more toward uniformity, though there is by no means a single plumbing code. Organizations are at work encouraging the adoption of a uniform, nationwide plumbing code.

The most obvious variations in codes have to do with climate—pipes in Minneapolis have to be a lot farther underground (to prevent freezing) than they do in Phoenix. Then, too, plastic piping, still a relative newcomer in the industry, is not accepted uniformly throughout the country. Regulations concerning it may vary widely; some places don't allow its use at all. But improvements in the materials and an increasing amount of experience with plastic piping seem to indicate that more uniform regulations can be expected. Since changes will always occur—in materials, in methods of joining pipes—an absolutely current plumbing code will never be possible.

Every home plumber should have a copy of the local code. (Get this from the building inspector's office; cost is usually low.) When you do any plumbing, check the code first. If the material you're planning to use isn't mentioned in the code, check with the building inspector.

Some plumbing codes prevent anyone but a licensed plumber from doing certain kinds of work. Most codes, though, allow homeowners to do all plumbing work in homes they own and live in. Check your code before you begin any work.

Tools to keep handy

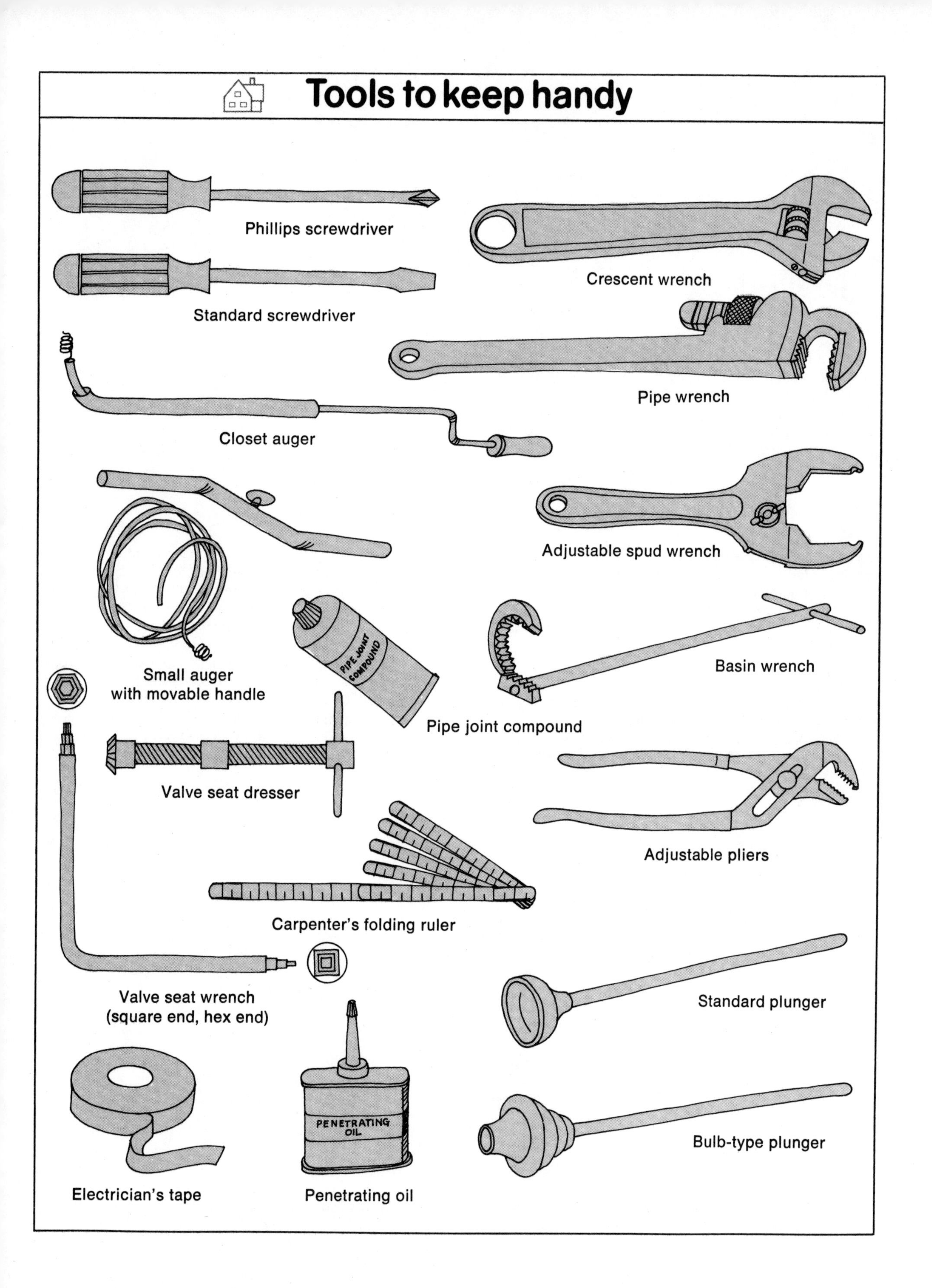

Phillips screwdriver

Standard screwdriver

Crescent wrench

Pipe wrench

Closet auger

Adjustable spud wrench

Small auger with movable handle

Basin wrench

Pipe joint compound

Valve seat dresser

Adjustable pliers

Carpenter's folding ruler

Valve seat wrench (square end, hex end)

Standard plunger

Bulb-type plunger

Electrician's tape

Penetrating oil

How-to-fix-it guide

When you find a leaking faucet or a clogged drain, making the repair yourself will save you money and spare you the inconvenience of waiting for a plumber. Knowing where to shut off the water supply (see page 4) and having a few basic tools on hand for the most common problems (see page 7) will give you a head start.

To repair your own fixtures, you'll have to buy replacement parts occasionally. Precise fit is always important, so when ordering parts, give the brand name of the fixture and the exact size of the part. Parts for older fixtures may not always be available, but accurate and detailed information will help your supplier find a part that will do the job.

HOW TO FIX A LEAKING FAUCET

The sooner you fix a leaking faucet, the faster you cut down on water waste—and those small drips can add up to a tremendous amount of water over a period of time. The method you use for fixing the faucet will depend on whether it has one or two handles for hot and cold-water control.

Leaking two-handle faucet

A faucet with one control for hot water and one for cold can either be a compression or washerless type. The washerless faucet substitutes diaphragms and discs for the washers and valve seats found on compression faucets.

Compression faucet. At the end of the stem of a compression faucet, you'll find a washer held in place by a screw. Ideally, when the faucet is turned off, the stem is screwed all the way down, and the washer fits snugly into the valve seat, stopping the flow of water. If the faucet is dripping from the end of the spigot, it could mean that either a washer or a valve seat has deteriorated. If the leak comes from around the stem, you'll need to replace the packing.

If the faucet "chatters" when turned on and off, a loose washer is the culprit—you could just tighten the screw that holds the washer in place. But with the faucet disassembled, you might as well install a new washer.

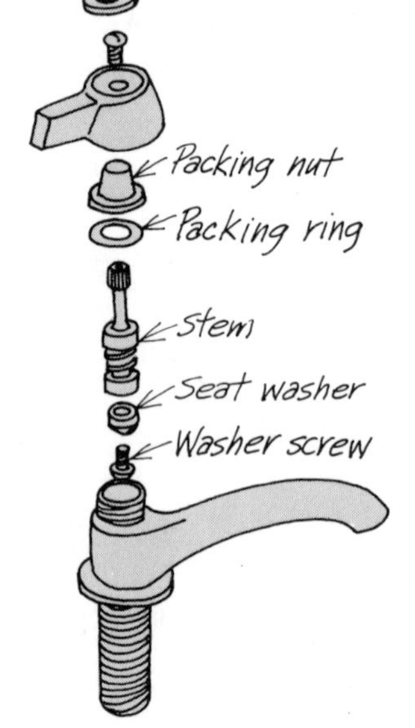

Compression faucet *is most common of two-handle types. It's easy to take apart; all parts are replaceable.*

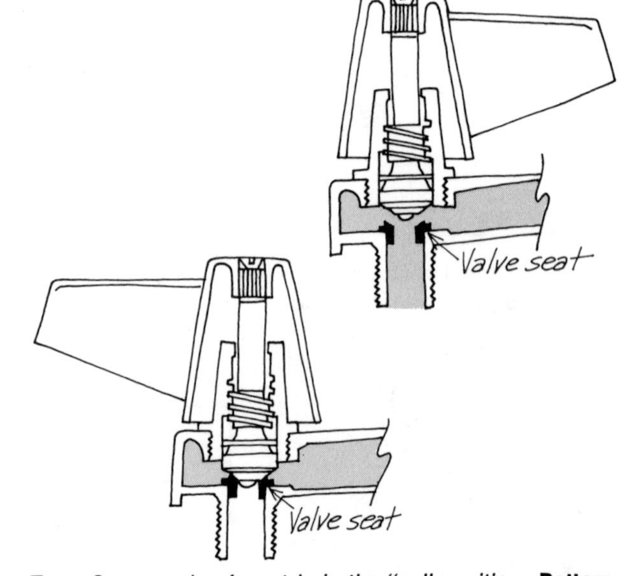

Top: *Compression faucet is in the "on" position.* **Bottom:** *Washer fits snugly onto valve seat when faucet is off.*

• *Replacing a washer.* Before you do any work on the faucet, turn off the water at the shutoff valve under the fixture. If you can't find a valve there, use the main shutoff valve in the basement or just outside the house where the water supply enters. Then take these steps:

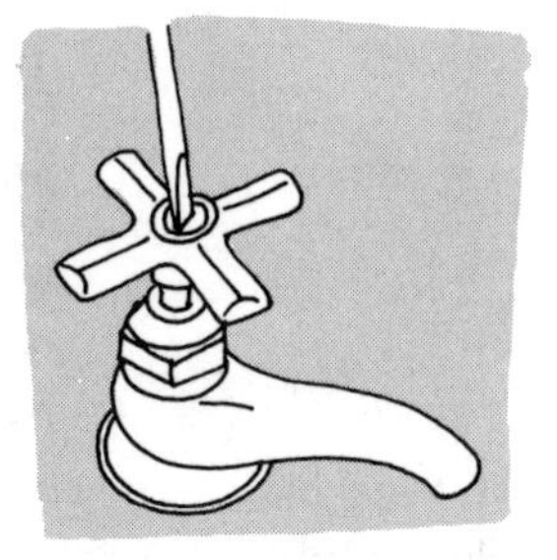

Step 1. *Remove the decorative insert that snaps or screws into the handle top. Then remove the faucet handle by unscrewing the handle screw that is now exposed.*

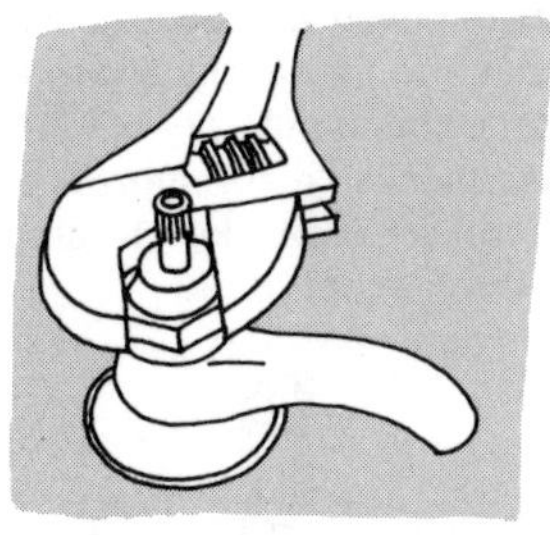

Step 2. *Remove packing nut with a wrench. (If the wrench has teeth, cover them with electrician's tape to prevent the teeth from scratching the finish on the nut.)*

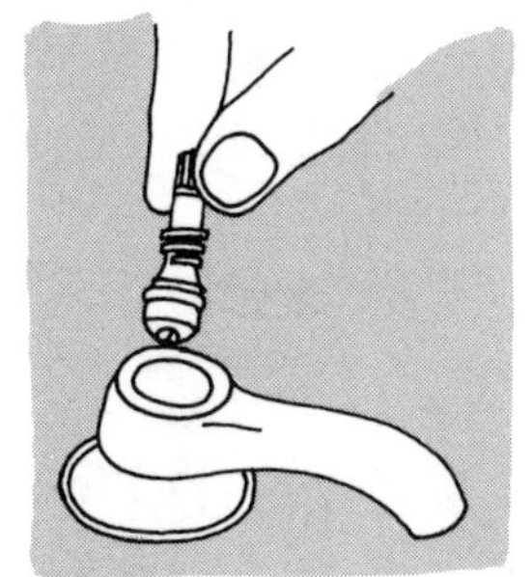

Step 3. *Unscrew the stem by turning counterclockwise (you may need a wrench to get it started). Take the stem all the way out and replace the washer with an identical one.*

Since washers may be flat or beveled (conical), be sure the replacement washer is not only the same size but also the same shape as the old one. If the exact washer isn't available, you can get a swivel-headed washer with a fitting that snaps into the threaded hole in the bottom of the stem.

If you break off the shank of a washer screw in its hole, you can drill a hole in the end of the stem and insert a swivel-headed washer. If you don't want to take the trouble to do this, replace the entire stem—all you need to do is get a new one the same size and screw it in.

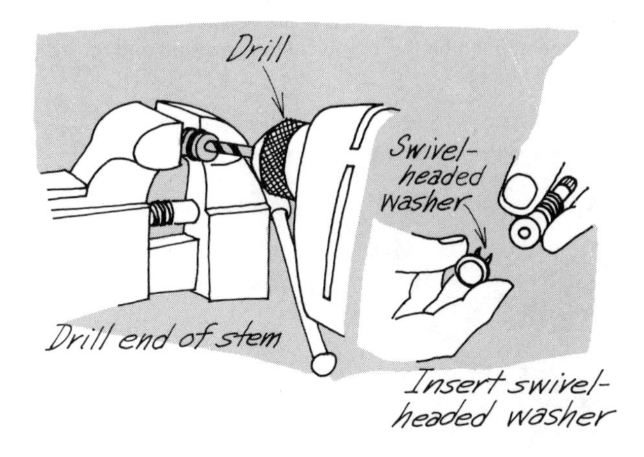

When you've put in the new washer, replace the stem, screwing clockwise by hand (lubricate the threads of the stem with a little grease or petroleum jelly). Replace the packing nut, tightening it with a wrench; replace the handle and screw in handle screw. Turn the water back on and check to see if dripping persists. If not, snap or screw any decorative cap back into place. If the dripping hasn't stopped, a damaged valve seat is the problem.

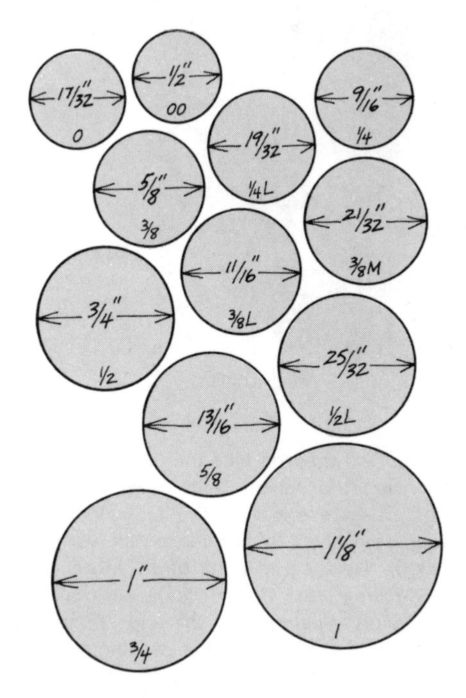

The chart above *shows the actual sizes of washers. Use it to measure the washer you need to replace. It's a good idea to have on hand washers of various sizes. You can buy assorted sizes of washers in a kit for this purpose.*

• *Repairing a valve seat.* Most compression faucets have a replaceable valve seat. To remove the old one, you'll need an allen wrench or a valve seat wrench. Disassemble the faucet as described previously. Then insert the wrench down into the faucet body and fit it into the valve seat. Remove the seat by turning counterclockwise. Take the seat to a plumbing supply store or hardware store and buy one identical to it. Install it, using the wrench; put some pipe joint compound on the threads of the new seat and reassemble the faucet.

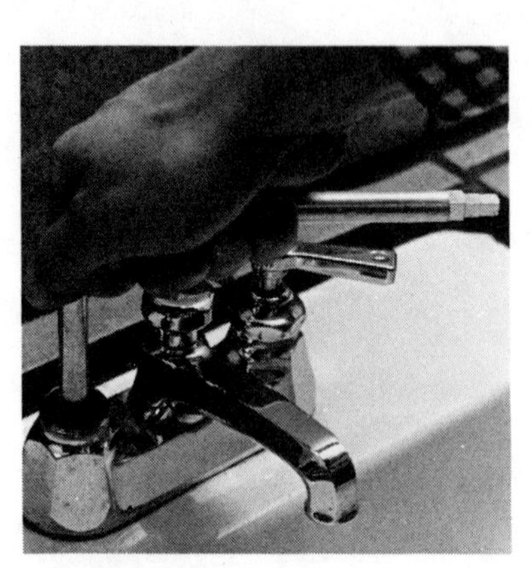

Remove valve seat *with allen or valve seat wrench.*

If the valve seat is not the replaceable type, use a simple, inexpensive tool called a "seat dresser" to grind down any burrs the valve seat might have. This makes it smooth and level again.

Use seat dresser *to grind burrs from valve seat.*

To use the seat dresser, take the faucet apart and fit the tool down onto the valve. Push down with a moderate amount of pressure and turn the tool clockwise several times. Clean out chips by blowing on the valve seat or by using a cloth. Replace all parts and turn on the water.

• *Fixing a leaking stem.* If water leaks out from around the stem instead of dripping from the end of the spigot, the problem is the packing inside the packing nut.

Turn off the water and disassemble the faucet. Some faucets use packing rings or rubber or plastic O-rings. Replace the ring with one exactly the same size and type. If your faucet doesn't have a ring, use graphite packing. One type is like string; wrap it a few turns around the spindle. The other kind is like soft wire; wrap it once around the spindle. Tighten the packing nut.

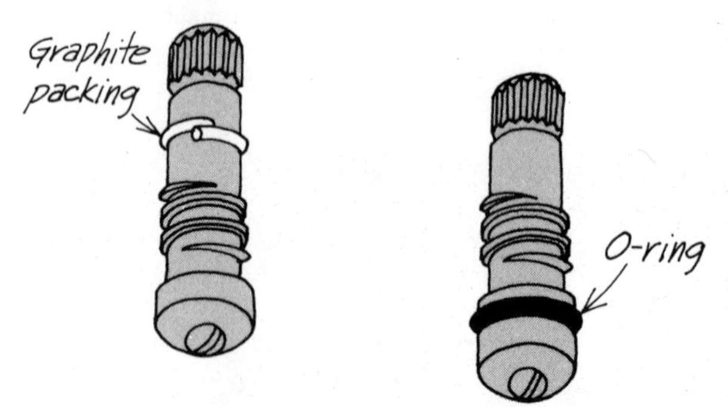

Washerless faucet. Metal discs or rubber diaphragms control the flow of water in washerless faucets. A leak in one of these faucets means some wear or damage to the diaphragm assembly, which is easily removed and replaced.

To remove a faulty diaphragm, first turn off the water supply at the shutoff valve. Then unscrew the handle and lift it off (having first removed any decorative insert). Now unscrew the stem nut and lift out the entire mechanism. If the diaphragm doesn't come out, pry it out carefully with a screwdriver. Take the whole assembly to a plumbing supply store to get the correct repair kit. Replace all faulty parts and reassemble the faucet.

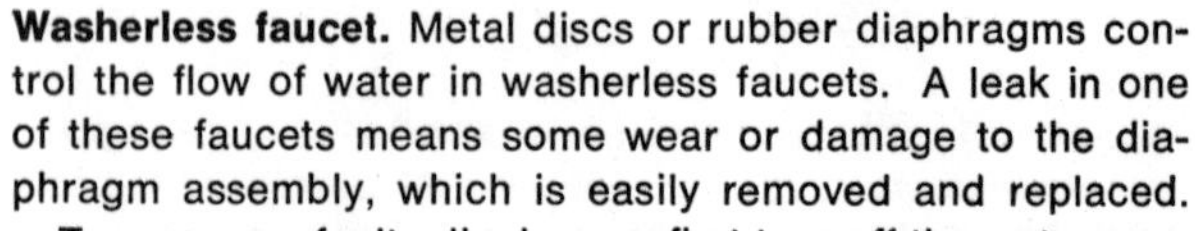

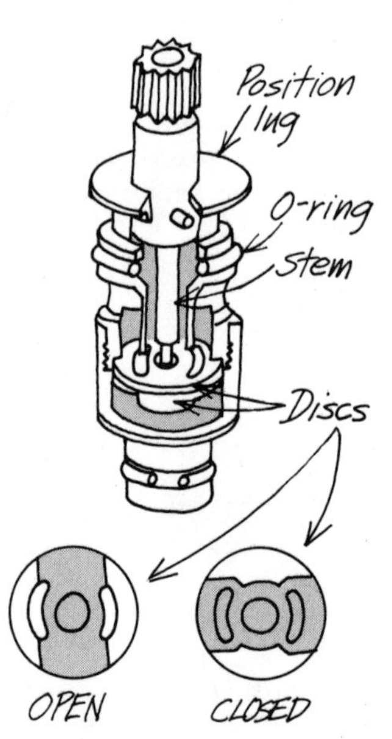

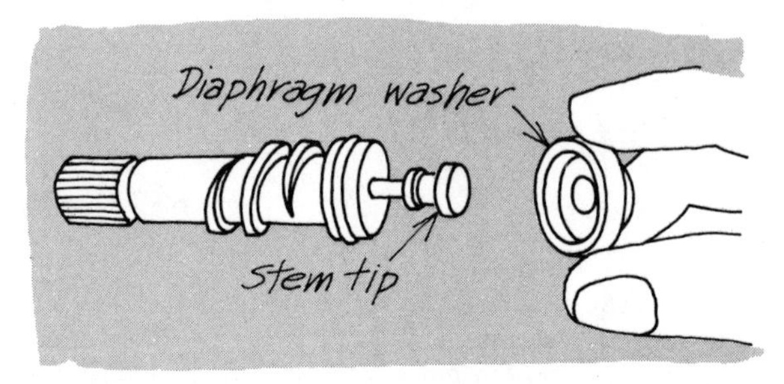

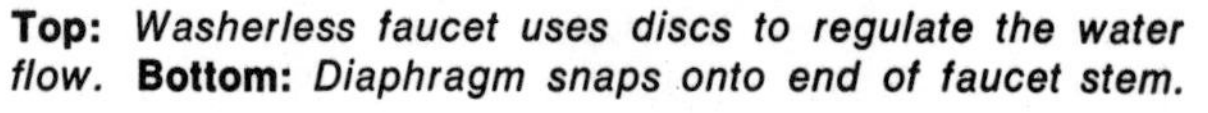

Top: *Washerless faucet uses discs to regulate the water flow.* **Bottom:** *Diaphragm snaps onto end of faucet stem.*

Leaking single-handle faucet

This type of faucet controls both hot and cold-water flow with one handle. It works by aligning interior openings with hot and cold-water inlets. The five basic types of single-handle faucets are pictured below and on the next page. Details will vary from brand to brand. Determine whether yours has a tipping valve, a cartridge, disc, or a brass ball to regulate water flow, and follow the instructions for repairing that type.

Tipping-valve faucet. The handle of a tipping-valve faucet operates a cam in the center of the faucet body. The cam, in turn, tips valve stems, allowing water to pass through the valves and up and out the spout. Dripping at the spout is caused by failure of the valve stem assembly or of the valve seat.

All parts of the valve assembly are replaceable—you can buy a replacement kit at a plumbing supply store. To get to the inner workings, first remove the spout by unscrewing it at its base. Then remove the decorative escutcheon. Use a wrench to unscrew the strainer plugs on either side of the faucet body. Remove gasket, strainer, spring, and valve stem by hand. Remove the valve seat with a valve seat wrench. Replace all old parts with new ones from the kit. (Put some pipe joint compound on the threads of the valve seat before installing it.)

Leaking at the base of the spout is due to the failure of the O-ring. Remove the old O-ring and replace it with an identical new one.

If flow from this faucet is sluggish, it is likely the strainers are clogged with sediment from hard water. Remove the plugs and gaskets and pull out the strainers. Rinse them out thoroughly and reinstall.

Disc-type faucet. Another type of single-handle faucet relies on a set of two discs to mix hot and cold-water flow. When lifted by the faucet handle, the top disc controls the amount of flow, and when rotated by the handle, it controls the mix of hot and cold water.

Leaking in this kind of faucet means the rubber seals are worn. To replace them you'll need to take the faucet apart. Remove the handle and decorative escutcheon. Unscrew the two screws that hold on the top of the cartridge and lift it off. You may have to pry out the stationary disc; do this slowly and carefully. Under the stationary disc is the seals, which you should take with you to a plumbing supply store in order to get the correct repair kit. Replace the seals and reassemble the faucet.

Cartridge faucet. When a central cartridge controls the water flow, a leak can be fixed only by replacement of the entire cartridge. Sometimes hardware and plumbing supply stores won't have the part in stock; you may have to order it from the manufacturer.

To install the replacement cartridge, first remove the faucet handle screw — it's probably concealed under a decorative inset—and lift off the handle. If there is also a decorative escutcheon below the handle, remove that too.

Then remove the stop tube, exposing the faucet stem. The stem is part of the cartridge, which is held to the faucet body by a clip. On some faucets the clip fits into place from the outside; on others the clip is under the handle. Simply pull out the clip and the cartridge.

When you've replaced the cartridge, take care to reassemble the faucet correctly. If there is a flat side on the stem, keep it pointing up during reassembling. Keep the pointer on the handle facing up, too.

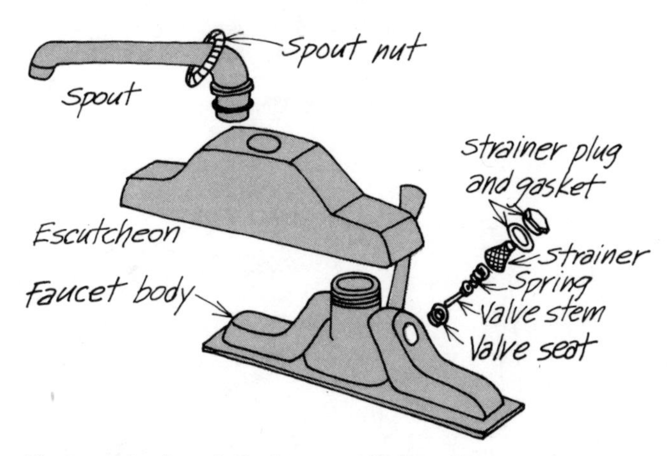

Tipping valve faucet. *Replacement kit has all parts.*

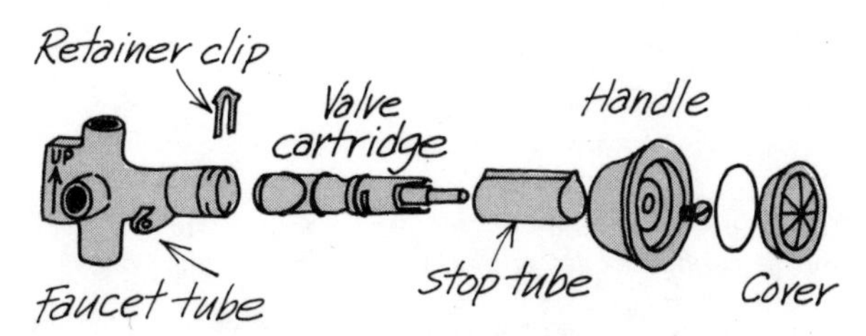

Cartridge faucet. *Cartridge is replaced as a unit.*

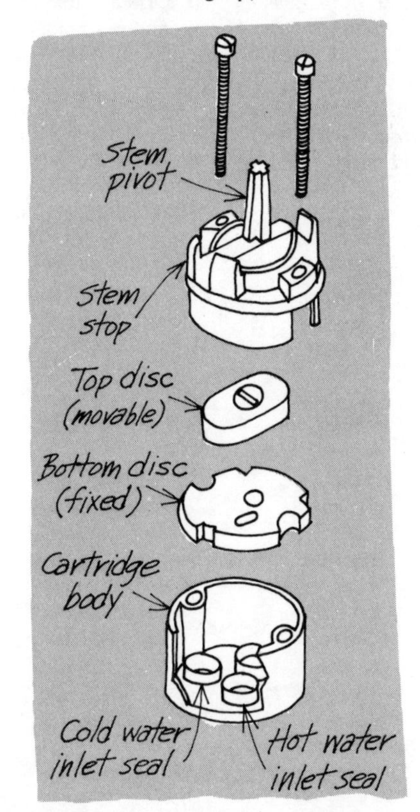

Disc-type faucet. *Movable top disc regulates water flow.*

Wall-mounted cartridge faucet. This has a central cartridge that can be completely replaced if the faucet leaks. After removing spout, handle, and escutcheon, remove the four mounting screws holding the assembly in place. Lift off the assembly and take it to a plumbing supply store for an exact replacement.

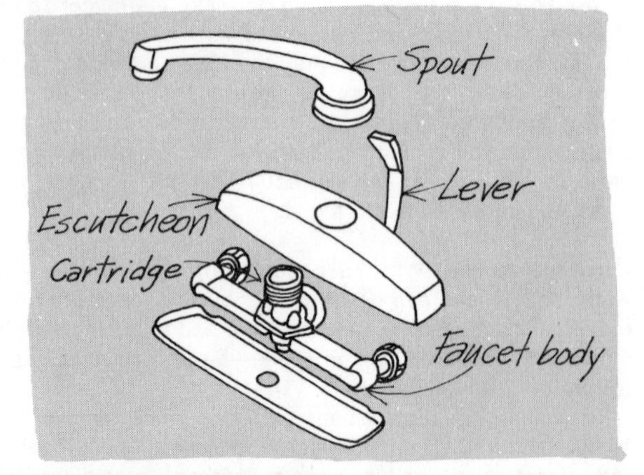

Wall-mounted cartridge faucet. *To correct leak, replace cartridge by unscrewing it from faucet body, lifting off.*

Brass ball faucet. Another type of single-handle faucet has a brass ball that regulates the flow and mix of hot and cold water. Openings in the ball align with hot and cold-water inlets in the faucet body to permit flow.

This type of faucet has gasket seats and springs at the inlets in the faucet body. If the faucet leaks, either a gasket or a spring is worn.

To replace worn gaskets and springs, first remove the handle by unscrewing the handle screw (concealed under a decorative inset, usually below the handle) and lifting it off. Pull out the brass ball by the stem. Underneath it are the springs and gaskets at the inlets. Take them out and get an identical set at a plumbing supply store. Install new gaskets and springs and reassemble the faucet.

Left. Brass ball faucet. *Leaks occur when the springs or rubber gaskets wear out. Remove brass ball to replace them.*

Leaking at the base of the spout

A faucet assembly with a swing spout may leak from the base of the spout where it joins the faucet body. The culprit in this case is a faulty washer or O-ring. Remove the spout by turning the nut at the base of the spout counterclockwise. Get an identical replacement for the old washer or O-ring and install it.

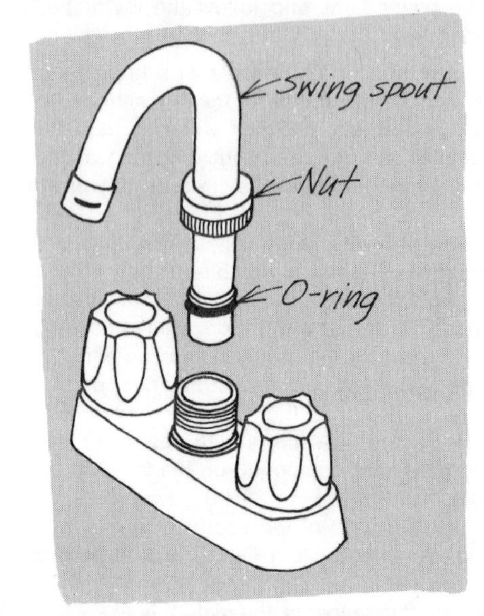

Swing spout *comes off faucet body when grooved nut is unscrewed. Replace the washer or O-ring to fix leak.*

Leaking tub and shower faucets

Faucets on tubs and showers may be either the two-handle (usually the compression type) or single-handle styles. If the handles are wall-mounted and the body is behind the wall, you'll need a socket wrench to unscrew the bonnet in order to get the stem out (see illustration below).

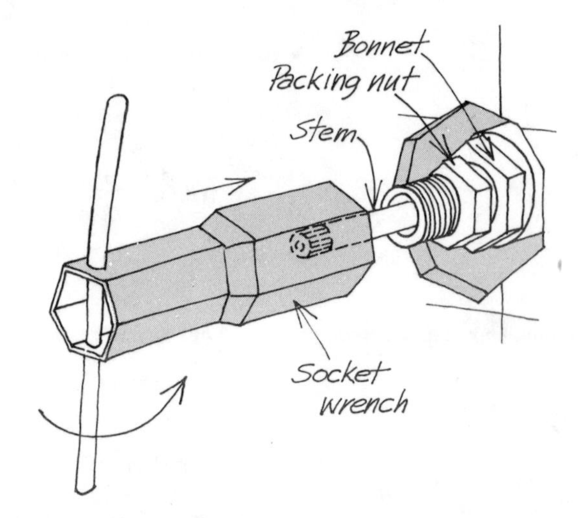

To remove *wall-mounted tub and shower faucets, use a socket wrench to get at the large nut behind the wall.*

HOW TO FIX A TOILET

The diagram below shows the inner workings of a conventional toilet. When you push the handle to flush, the trip lever pulls the tank ball out of the valve seat. Then the water in the tank flows rapidly downward through the flush valve outlet into the bowl or into the leg of the trap, causing water to overflow the trap. This overflowing creates a siphoning effect, pulling all the water in the bowl with it down the drain.

Inside the tank, as the water level descends, the tank ball falls back into place on the flush valve, and the float ball descends with the water level. The ball pulls the float arm down, an action which opens the float valve in the ballcock assembly. Open, this valve lets water into the tank through the filler tube. It also causes the empty bowl to be refilled by sending water through the refill tube and down through the overflow tube.

As the water in the tank rises, so does the float ball, pulling up the float arm. At a certain level, the ball will have raised the arm high enough to shut off the float valve, and the cycle will be complete.

If the water in the toilet is continuously running or if the fixture makes a whistling noise, the trouble is at the float valve or the flush valve. Both are inside the tank.

Running toilet

A continuously running toilet means the tank ball isn't seated properly in the valve seat at the bottom of the tank. The tank ball itself may be worn and need replacing. You can tell at a glance if it's rough and worn around the edges. If it is, unscrew it, take it to a plumbing supply store to make sure you get an exact duplicate, and screw the new one in. To keep the water from running while you're working on the toilet, turn it off at the valve under the tank or slip a stick under the float arm and prop it on the edges of the tank to keep the float valve closed.

If the tank ball isn't worn, the problem may be that the valve seat is rough, scaled, or corroded. If so, turn off the water inlet under the toilet, empty the tank by flushing it, and sponge out the remaining water. Dry the valve seat with a cloth. Sand the dried seat with emery paper until it's smooth again. Turn the water back on.

If the tank ball and valve seat show no signs of deterioration, check the tank ball guide wires. Take the top off the tank, flush the toilet, and watch the action of the guides. If the tank ball does not drop freely back into place as the water descends and if it doesn't fit snugly into the valve seat, the wires are probably misaligned or bent.

(Continued on next page)

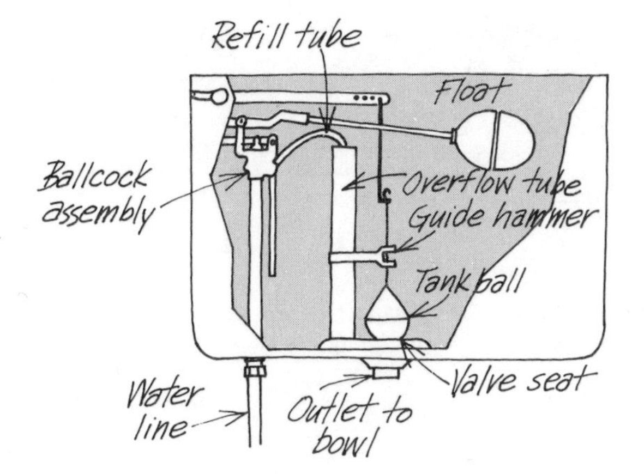

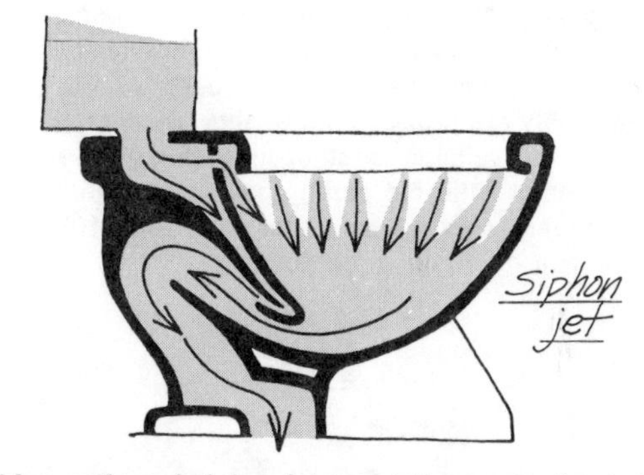

Flushing action *of siphon-jet type toilet is accelerated by a stream of water forced into leg of trap at rear.*

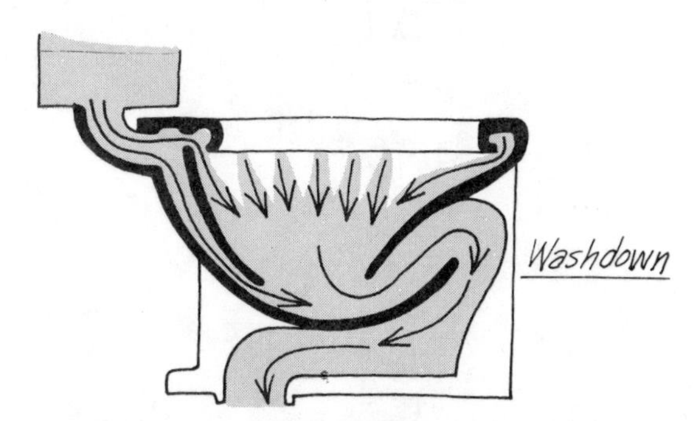

Water enters *washdown toilet from rear and from holes in rim. Water spills over leg of trap, causing siphon action.*

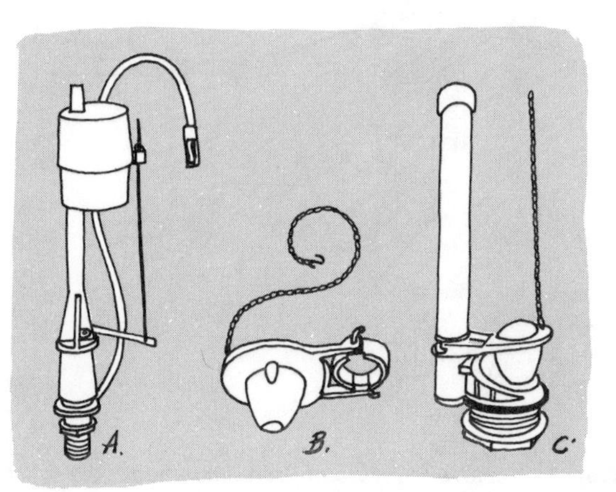

Left. *Three modern replacements for tank parts: A) Faster-filling ballcock assembly; B) Flapper replaces tank ball; C) Replaces tank ball and refill tube.*

If the guide hammer is not lined up directly above the valve seat, loosen the screws holding the hammer to the overflow tube and reposition it.

If the wires are bent, it may be easier to replace them than to straighten them. You'll find replacement wires at all hardware and plumbing supply stores. Check the way the old ones were installed before you remove them; then install the new ones exactly the same way.

Wire corrosion can be another problem; inspect for this. Dry corroded wires thoroughly and sand them smooth with emery paper.

Whistling toilet

If your toilet constantly makes a whistling or singing noise, the float valve inside the ballcock assembly is failing to close fully, causing water to run through it into the toilet, sometimes making that high-pitched sound.

Lift up the float ball. If the running and whistling stop, the float ball simply isn't rising enough to shut the float valve entirely. If that's the case, the water level in the tank will be too high, and water will be flowing over the top of the overflow tube down into the bowl. The tank water level should be 1 or 2 inches below the top of the overflow tube.

Bend the float arm downward to cause the ball to shut off the float valve sooner. It may be that the ball is partially filled with water and can't rise high enough to shut off the float valve. Unscrew it and replace it with a new one. A plastic float ball won't corrode the way copper ones do.

If adjustment of the float ball and arms doesn't stop the whistling, the float valve itself probably has faulty washers or a damaged valve seat. Separate parts within the ballcock assembly are replaceable; washer, plunger, and valve seat are often available together in a kit. Disassemble the ballcock, remove the old parts, and reassemble. You can use a standard seat-dressing tool (see page 10) to grind the valve seat if you don't replace it.

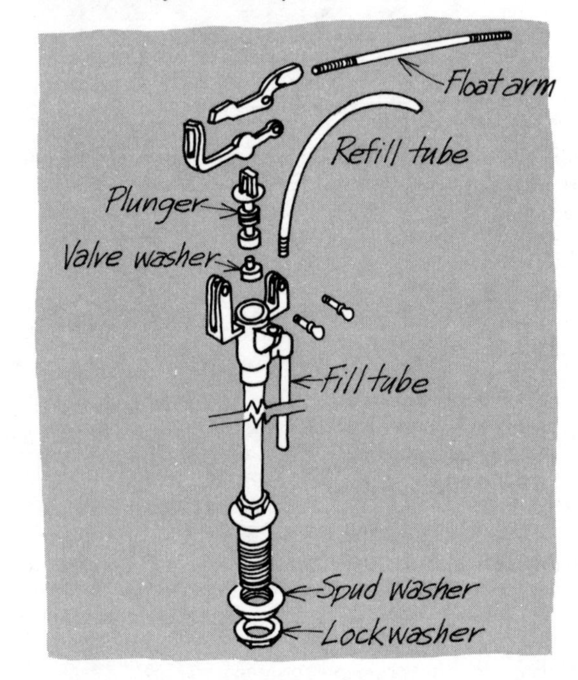

Float arm *lifts plunger and valve washer up off valve seat; water flows out fill and refill tubes into toilet.*

If the entire ballcock assembly looks as if it needs replacing, take the following steps:

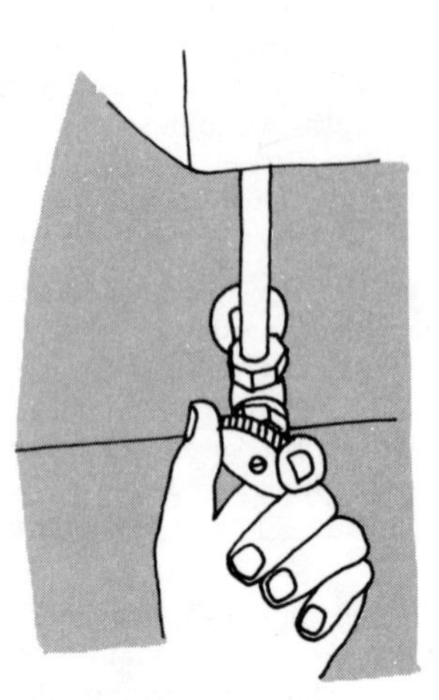

Step 1. *Shut off the water. Drain and dry the tank.*

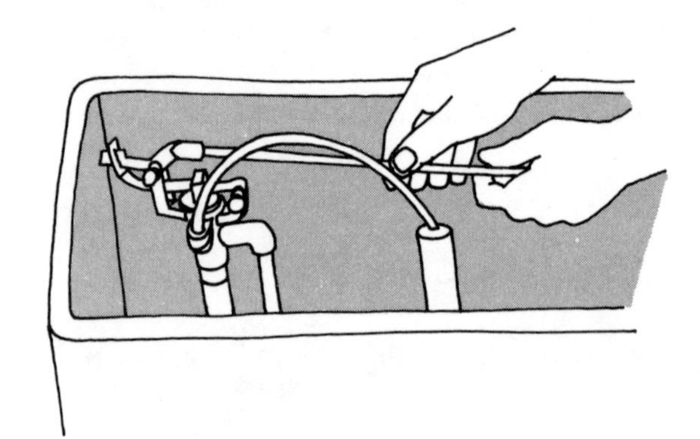

Step 2. *Unscrew the float ball arm from the ballcock (and save the ball and arm if they are sound).*

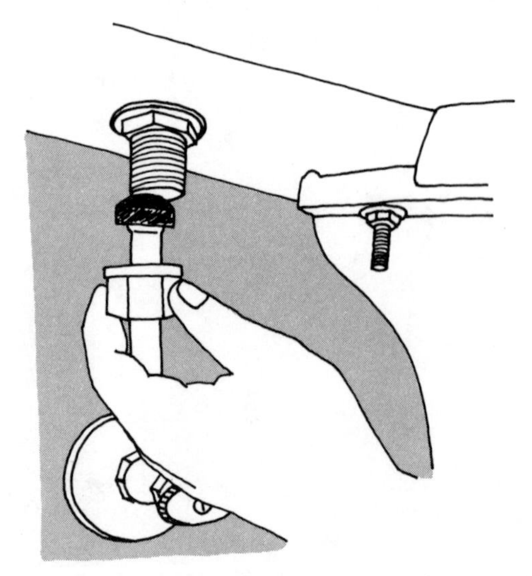

Step 3. *Remove supply pipe from underneath the toilet.*

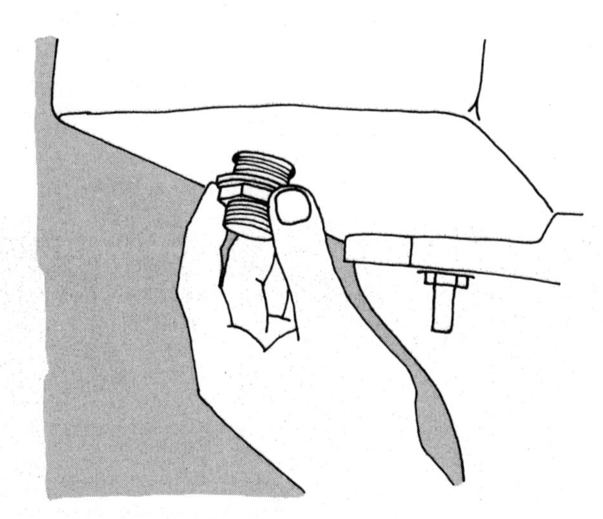

Step 4. *Working on the underside of the tank, loosen and remove the locknut and washer.*

Step 5. *Remove the ballcock assembly from the tank.*

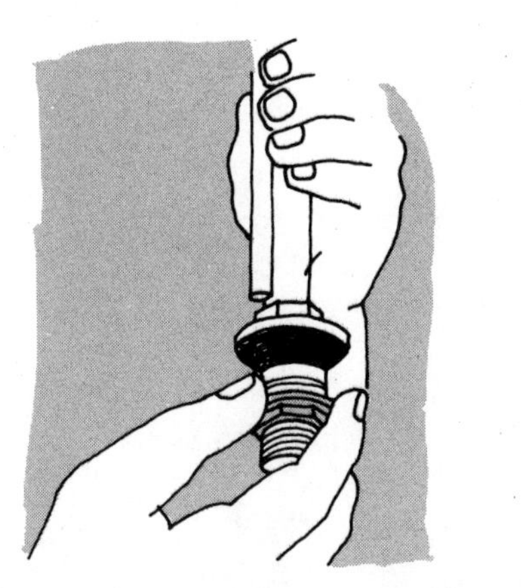

Step 6. *Remove the coupling nuts, locknuts, and washer from the shank of the new unit; fit the unit into the tank. Secure the unit loosely with washer, locknuts, and coupling nuts. Position the unit so that the refill tube can be clipped in place with its mouth over the overflow tube and so that the float ball can rise and fall freely in the tank.*

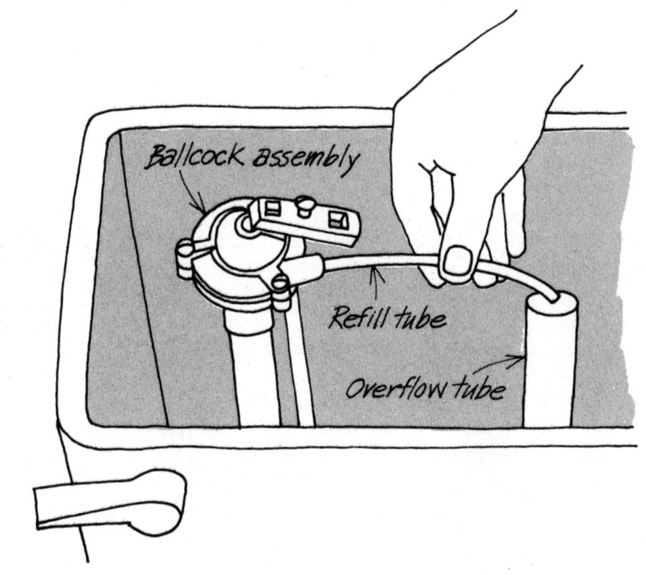

Step 7. *Adjust the refill tube into the overflow tube.*

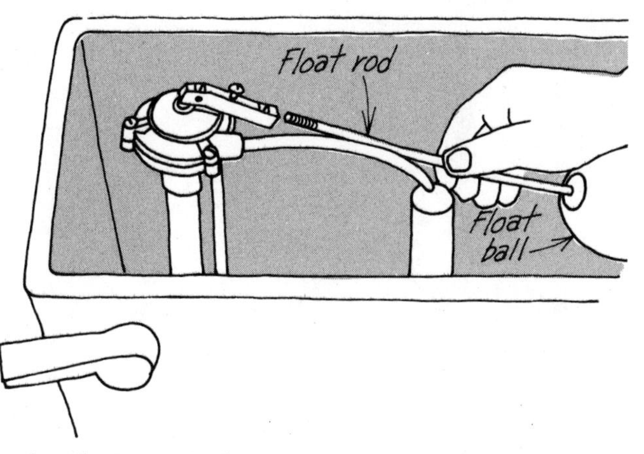

Step 8. *Attach the float ball arm to the ballcock in the tank. Then, tighten the lock and coupling nuts firmly.*

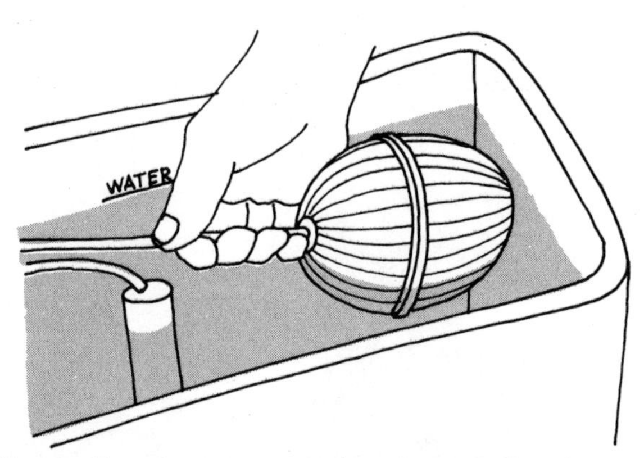

Step 9. *Turn the water on at the valve and allow the tank to fill. Adjust the float ball arm if necessary to obtain proper water level in full tank. (Water level should be about 2 inches below the top of the overflow tube.)*

Leaks at the tank's base

A one-piece toilet, of course, won't leak between tank and bowl because the two comprise a single unit.

But a two-piece toilet may leak at the base of its tank, where the ballcock assembly connects with the water supply pipe or at the tank outlet.

The tank and bowl of two-piece toilets connect one of two ways: 1) a wall-hung tank connects with a short, 90° bend of brass pipe; 2) a bowl-mounted tank fits directly onto the back of the bowl and is held in place by two bolts.

Before you begin repair work, turn off the water supply and drain and dry the tank.

When fixing a leak between tank and bowl of a wall-hung toilet, you'll have to disconnect the pipe between the two and lift the tank off its wall bracket. To remove the pipe, unscrew the nuts at both ends with a spud wrench. Then lift the tank off the bracket. Replace all washers at both connections. Check for excessive wear or damage at the flush valve and replace it if necessary. Reassemble the toilet by fitting the pipe into the bowl first and then fitting the tank down onto the pipe and the wall bracket.

A leak at the outlet of the bowl-mounted toilet means damage in the flush valve assembly or a worn rubber washer where the flush valve pipe enters the bowl.

Turn off the water, drain and dry the tank, and remove the guide hammer, lift wires, and flush ball. Detach the refill tube from the overflow tube and move it aside. Remove the two bolts holding the tank to the bowl and lift off the bowl. Underneath is a locknut that holds the flush valve tight. Remove that nut, pull out the flush valve, and check for damage to the valve or washers. If the valve is corroded or damaged, buy an exact replacement and install it.

Look at the large, conical rubber washer from the point where the flush valve enters the top of the bowl. Replace it with a duplicate if it is worn or damaged. Reassemble the tank.

To replace a leaking washer on the intake pipe, first turn off the water and empty the tank. Loosen the locknut and coupling nut. Lift the ballcock assembly just enough to free the washer if it doesn't slip out easily. Replace the washer and resecure the unit.

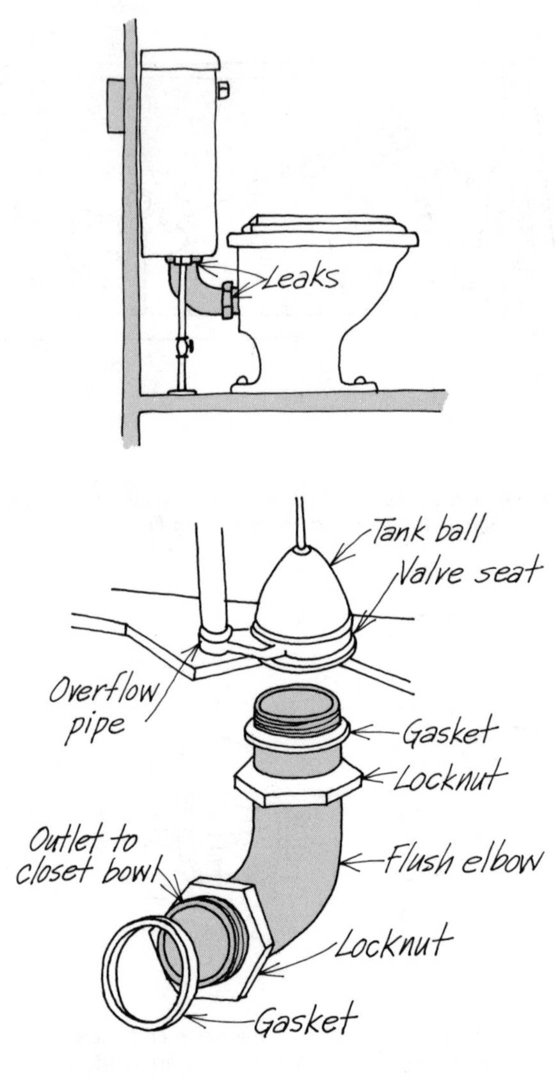

Wall-hung tank *can leak at either end of pipe because of failure of large rubber gaskets or flush valve seat.*

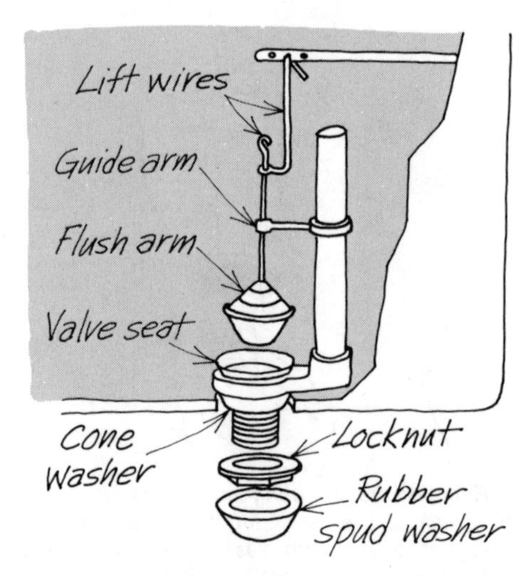

Bowl-mounted tank *will leak at the base when the flush valve seat is damaged or when the rubber gasket fails.*

Leaks at the toilet bowl's base

Leaking at the base of the toilet bowl is caused by the failure of the large gasket between the bowl and the toilet drainpipe. To replace the gasket, you have to remove the bowl—not an easy job. If you plan to do it, read this section first and then proceed carefully. First turn off the water.

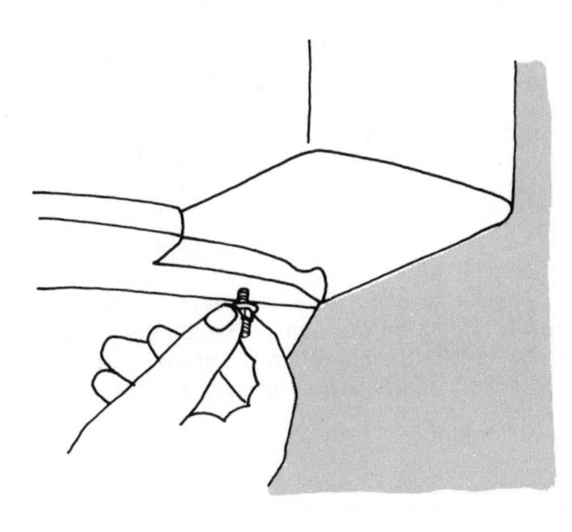

Step 1. *If it's a two-piece toilet, disconnect the tank from the bowl (see opposite page for instructions).*

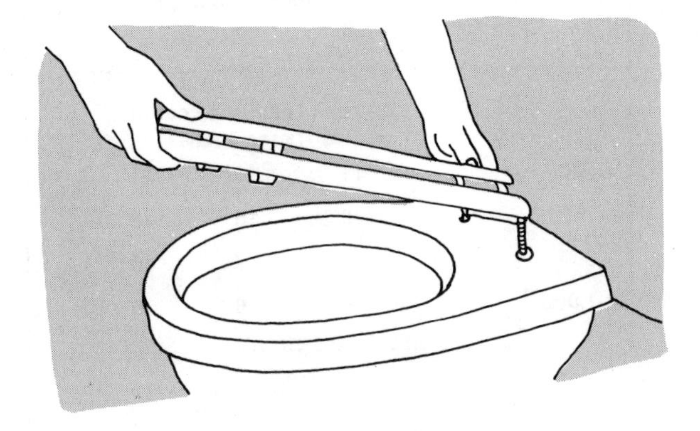

Step 2. *Take the seat and seat cover off the bowl.*

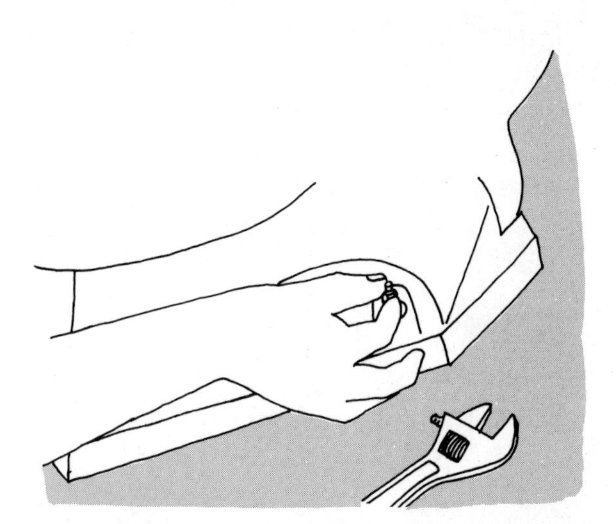

Step 3. *Remove the porcelain or chrome-plated bolt covers from the base of the toilet bowl. Unscrew the nuts.*

Step 4. *Break the seal between the bowl and the floor by jarring the bowl free. Then, lift off the bowl.*

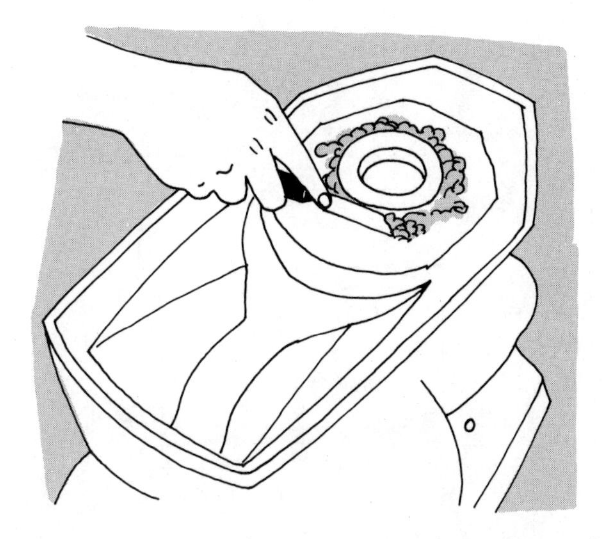

Step 5. *Remove the old gasket. Clean off the bottom of the bowl. Clean and dry the floor under the bowl.*

(Continued on next page)

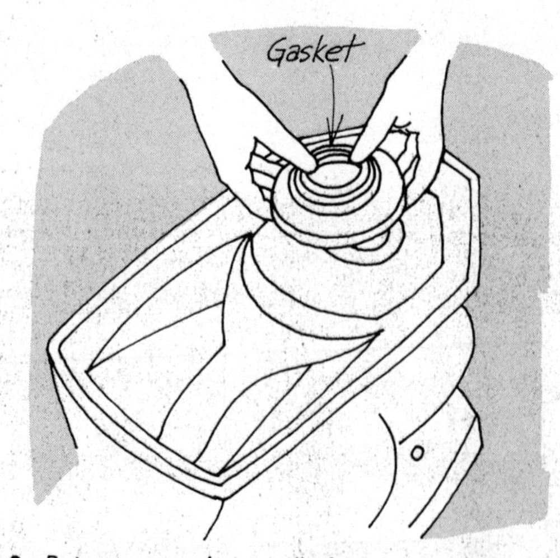

Step 6. *Put a new gasket on the base of the bowl. Apply a liberal amount of plumber's putty around the bottom rim of the bowl. Set the toilet down on the floor flange so the bolts in the floor go straight through the holes in the base.*

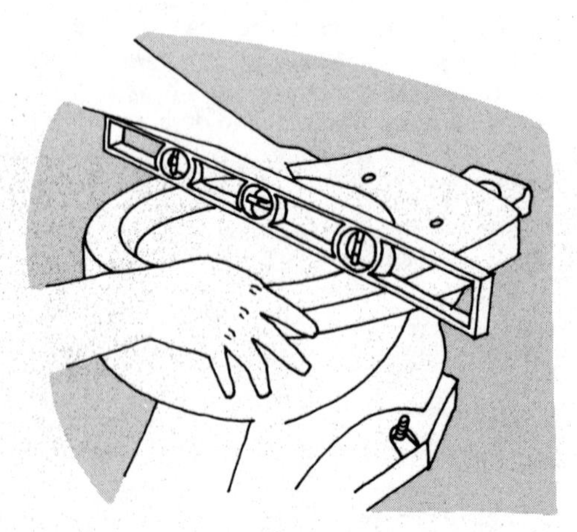

Step 7. *Use a level to assure a level surface—you may need to use shims on the bolts. Tighten the nuts carefully so you don't crack the porcelain. Replace bolt covers, toilet seat and cover. Replace the tank and turn on the water.*

Septic tank care

Like other parts of the house plumbing system, a good septic tank system doesn't require a great deal of maintenance. But the maintenance it does require is crucial since failure of the system can constitute a serious health hazard. You should have a diagram of your septic tank's layout, showing the location of the tank, pipes, manholes, and drainage field, to use when troubleshooting.

Pumping out the sludge that accumulates at the bottom of the tank is a maintenance task that must be done regularly. With normal use, a tank needs to be cleaned out every two or three years. It's worth the cost to hire a professional septic tank cleaner for this job. You can measure the sludge yourself, though, to determine when to clean the tank.

To measure the sludge, you need to know the capacity of your tank and the distance from the water line to the bottom of the tank (A). Measure the distance from the top of the sludge to the outlet pipe (B). (A towel tied onto the stick will tell you the sludge depth; a nail on the end of the stick will help you locate the bottom of the pipe.) If that depth is less than the figure given on the chart, it's time to have your tank cleaned.

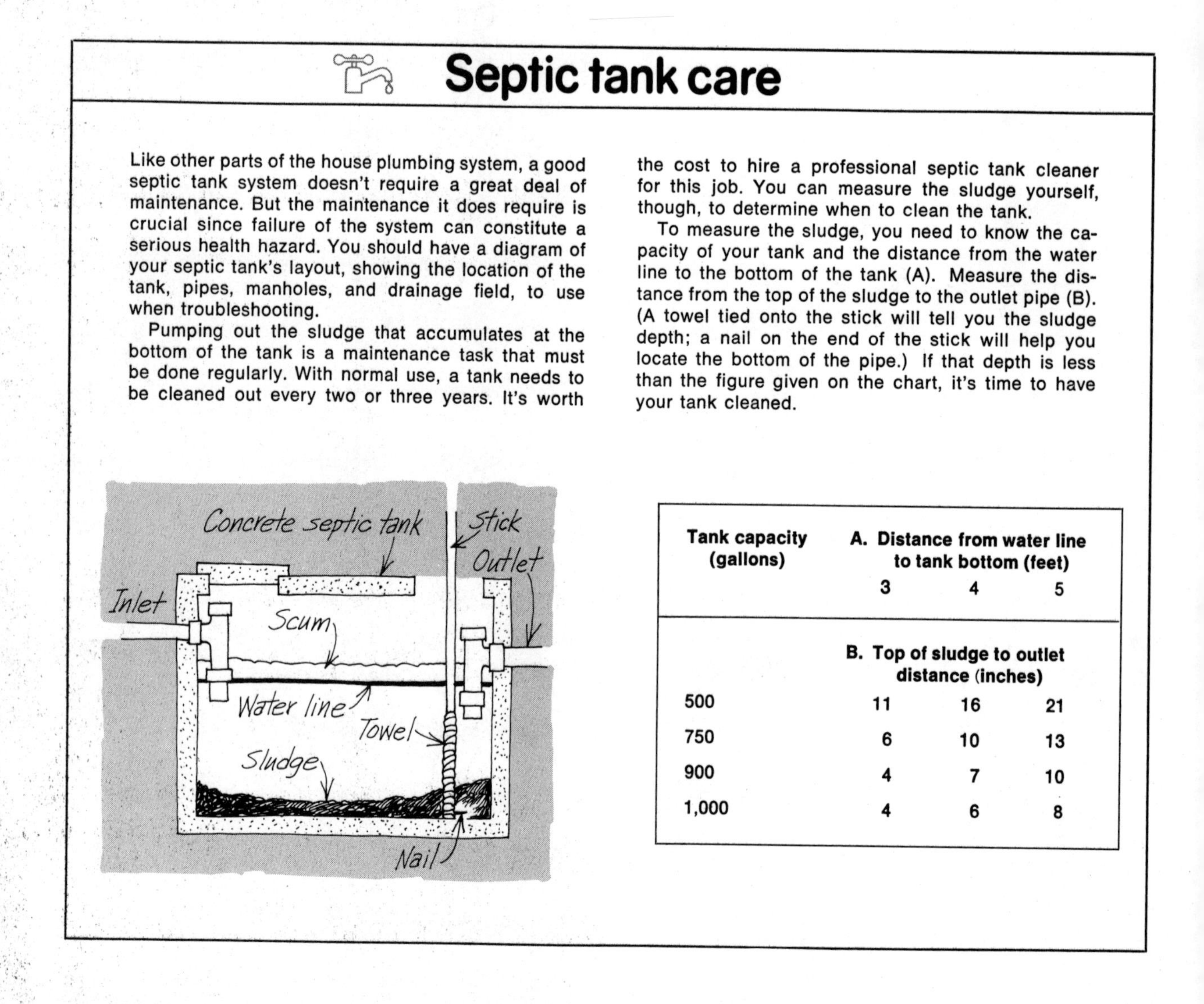

Tank capacity (gallons)	A. Distance from water line to tank bottom (feet)		
	3	4	5
	B. Top of sludge to outlet distance (inches)		
500	11	16	21
750	6	10	13
900	4	7	10
1,000	4	6	8

Sweating toilet tank

The cause of a sweating toilet tank is the same as the cause of sweating pipes: condensation occurs when water in the tank cools the porcelain and the cool porcelain meets the warm air in the room. Plumbing supply stores have insulation kits available for lining the insides of the toilet tank to prevent condensation.

Also available are tempering valves, which allow a little hot water to be mixed in with the cold water in the tank.

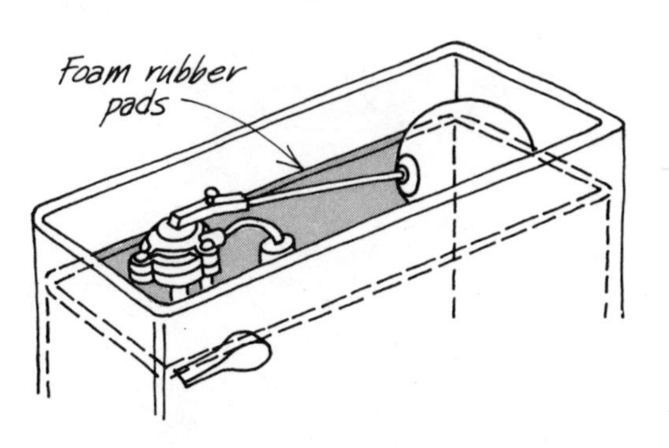

Foam jacket *fits inside the toilet tank, prevents sweating. You may have to order it from fixture manufacturer.*

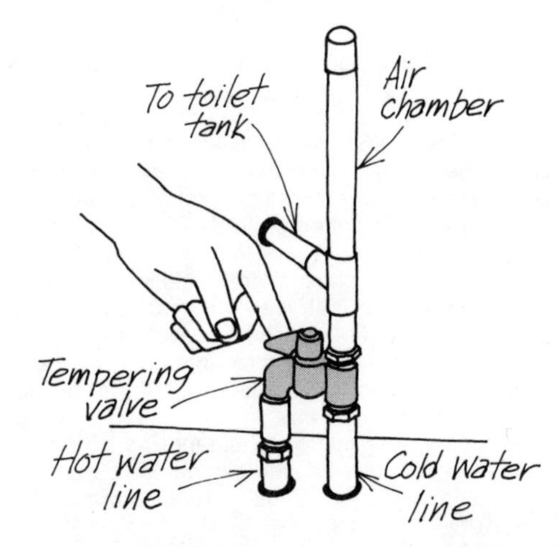

Tempering valve *permits a small amount of hot water to be released into the toilet tank, helps stop sweating.*

WHEN A DRAIN IS CLOGGED

No plumbing problem is more common or more frustrating than a clogged drain. Usually the problem is solved fairly easily and with little expense (though often with more than a little mess).

Clogged toilet

If the water in the toilet starts to rise above its normal level after flushing, you can usually prevent or stop overflow by quickly removing the tank lid and closing the flush valve by hand. When the water stops rising, look for an obstruction in the bowl outlet and try to dislodge it with a length of stiff wire—a straightened-out coat hanger will do. Don't use caustic drain cleaners in the toilet bowl: they aren't effective and will harm the porcelain finish.

If the toilet remains clogged, use a plunger that has a special tip to fit the toilet bowl. It creates a powerful force to either push the obstruction through the trap or draw it back into the bowl. Check to see if the stoppage is cleared by pouring water into the bowl. Flush only when the water level doesn't rise above normal with the addition of more water.

(Continued on next page)

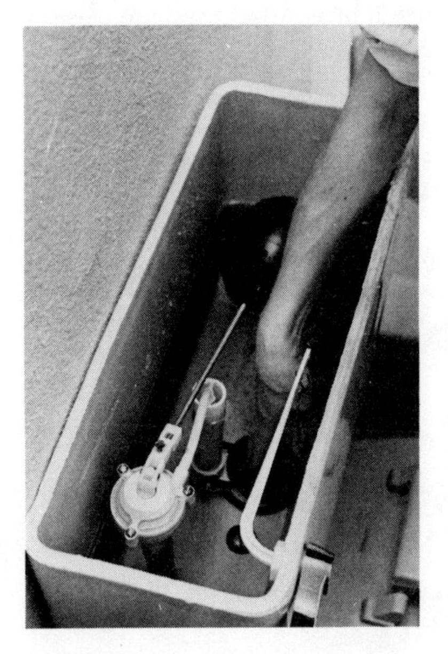

Stop overflow *by pushing down on tank ball (see page 13).*

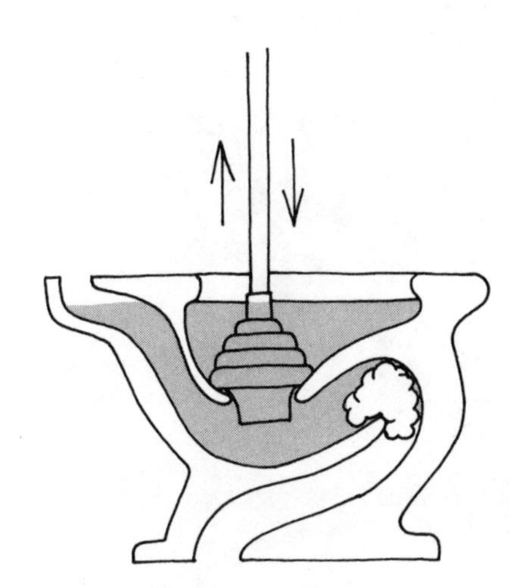

Bulb-type plunger *works especially well to clear clogged toilets.*

Preventing clogged drains

Most clogged drains can be avoided if you take a few precautions.

Kitchen sink drains clog most often because of the buildup of grease and the particles of food that get caught in the grease. Be especially careful to avoid letting grease go down the sink drain. Another real villain in the drain is coffee grounds—throw them out, don't wash them down.

Regular use of a caustic drain cleaner, whether the drain is clogged or not, is also a good precautionary habit to form. Follow the instructions on the package. You'll need to let the cleaner sit in the bend of the trap for awhile to be effective. Be careful not to splash it about or get any on your skin. Rinse the sink thoroughly after using the cleaner so no traces of it remain. Don't use caustic cleaner in a garbage disposer unless it's specifically recommended for that purpose.

Hair and soap are usually at fault when bathroom sink drains become clogged. Regular use of a caustic drain cleaner here too will help keep them unclogged. Many bathroom sinks have pop-up drains that can be lifted out; some require a half-turn to the left, others lift straight up and out. Hair collects in these drains, and it's a simple matter to lift them out once a week to remove the hair and rinse them off.

Occasional use of a drain cleaner in tub and shower drains also helps keep these drains unclogged. Use a cleaner less often here than in the kitchen sink. If the tub has a pop-up drain, lift it out occasionally, clean out hair and other matter, rinse and replace. Some tubs, showers, and basement floor drains have strainers held in place by one or two screws. You can remove these strainers and reach down into the drain to clear out all accumulated debris.

Don't use drain cleaner in the toilet bowl—it won't prevent clogs and will damage the bowl's finish.

Whenever you go up onto the roof of your house to clean out the drains and gutters with a garden hose, you can flush the DWV system by running the hose down into all vents and giving them a minute or two full flow.

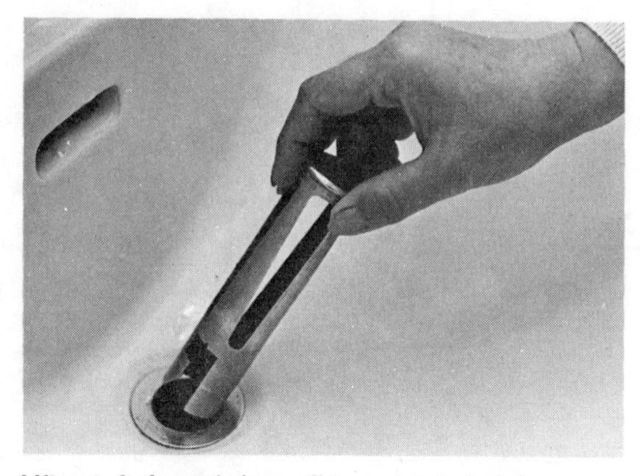

Lift out drain *and rinse off to prevent clogging.*

. . . Clogged toilet (cont'd.)

A closet (toilet) auger will allow you to reach down into the toilet if the plunger hasn't cleared the stoppage. A closet auger can negotiate the curves of the trap with a minimum of mess.

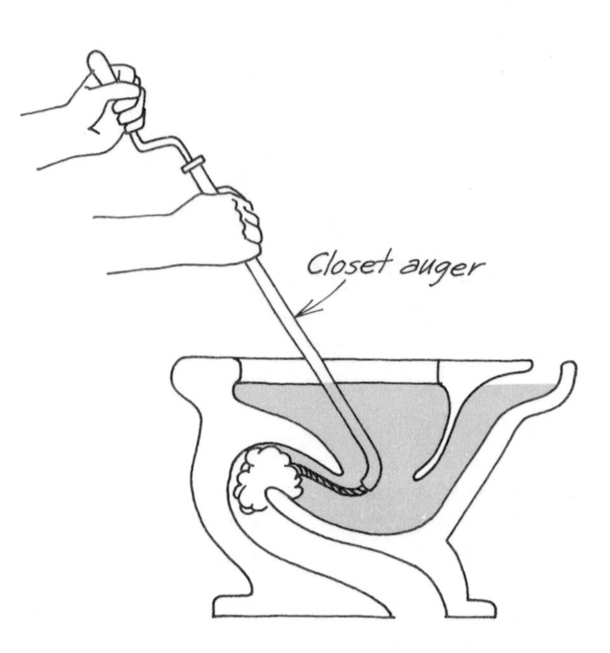

Closet auger *is designed specifically for negotiating the curve of the toilet trap. It's an inexpensive tool.*

Although you can use a regular auger, you'll need to force it into the trap by reaching down into the toilet. One useful technique in doing this is to pull a plastic garbage pail liner onto your arm and guide the tip of the auger down into the trap.

If neither the plunger nor the auger works to clear the obstruction, go to the nearest cleanout or vent and work an auger (or hose or balloon bag, as described on page 22) back toward the fixture. If other fixtures are also clogged, you'll need to clear the house drain (see page 23).

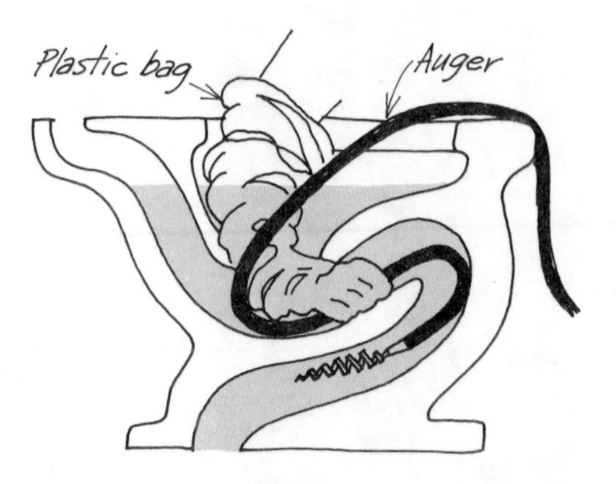

Conventional auger *can clear a stopped toilet, but first cover your arm with a plastic garbage bag to avoid the mess.*

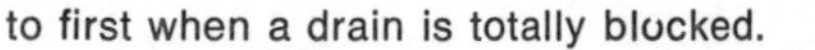

Slow-draining sink

Most blockages in sinks are caused by the buildup of grease, soap, and hair.

If the drain is only partially blocked—that is, if it will drain water away, but only very sluggishly—there are a couple of things to try.

Pour several kettlefuls of scalding water down the drain. This is quite likely to do the job, especially in a kitchen sink, where grease buildup is the probable cause of the clogging.

If the hot water doesn't work, you can try one of the commercially available caustic drain cleaners. These compounds work on such blockages as matted hair as well as grease. Caustic cleaners are very strong; use them with care. Read the instructions on the package and follow them exactly. Don't splash around. (If you get any on your skin, wash immediately with cold water.) Flush the drain thoroughly with hot water for several minutes after using.

Clogged sink drain

By whatever name you know it—plunger, force cup, plumber's friend, plumber's helper—it is the tool you should turn to first when a drain is totally blocked.

First, fill the clogged sink with water about halfway to the top. Coat the rim of the plunger with some petroleum jelly to insure a tight seal. Put the plunger down over the drain—slide the cup in at an angle so that no air remains trapped under it. If the fixture has an overflow drain, cover it with a wet cloth. To prevent loss of vacuum when working on a double sink, cover the other drain with a wet cloth. Now, work the plunger up and down vigorously a dozen or so times. If the stoppage clears up partially, you can use hot water or a caustic drain cleaner. If the plunger doesn't work at all, you'll need to try other methods detailed below.

Cleanout plug in trap. If the trap under the clogged sink has a cleanout plug, you have easy access to the probable location of the blockage.

Place a bucket directly underneath the plug. If you've used a chemical drain cleaner before this, be extremely careful when you open the plug. Remove the plug with a wrench, letting the water fall into the bucket. Use a piece of stiff wire (a straightened-out clothes hanger is good) and work it around in the trap; try to loosen and pull out the obstruction.

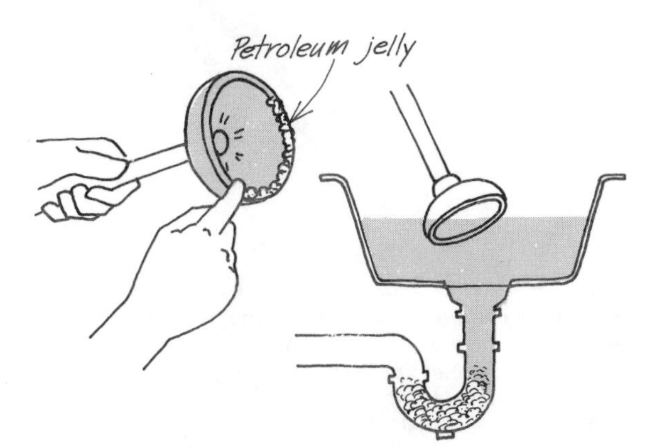

Coating rim *of plunger with petroleum jelly helps insure unbroken suction. Slide plunger into water at an angle.*

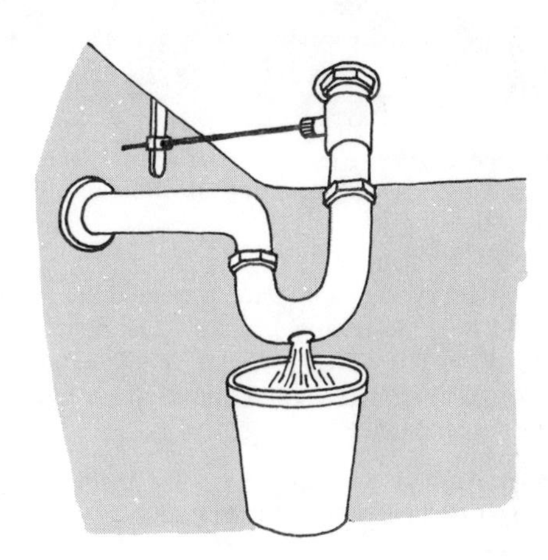

Place a bucket *underneath the trap before removing the cleanout plug. Use a stiff wire to break up the blockage.*

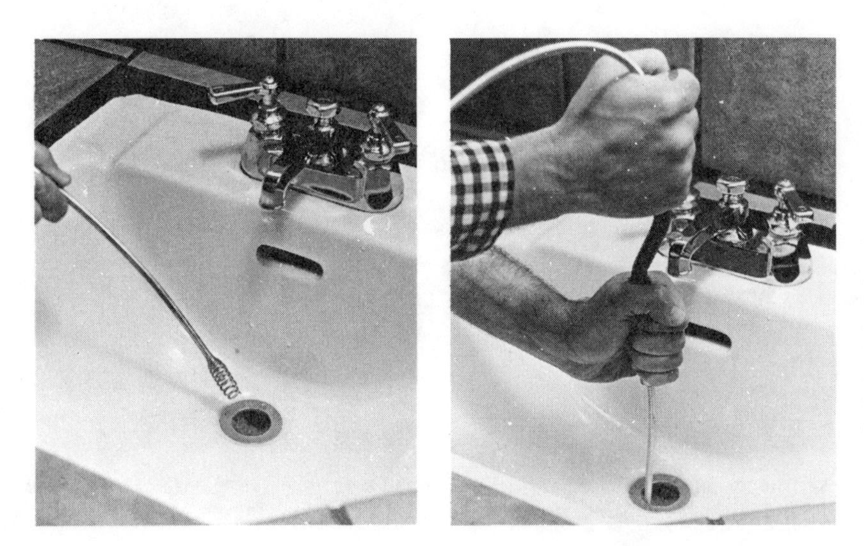

If the clogged sink *has no cleanout plug in the trap, or if you can't locate the blockage through the plug, try an auger or "snake." Feed it slowly down into the drain, turning constantly. When auger tip encounters blockage, force down hard while turning handle. Lift back out carefully.*

Removing and cleaning a trap. To remove a trap, first place a large pan or pail directly underneath it. Loosen the slip-nuts at the tops of both legs of the trap (protect the finish of chrome-plated pieces by covering the jaws of the wrench with electrician's tape). Ease the washers away from the joints so they don't split. Pull the trap downward to free it, spilling the contents into the pan.

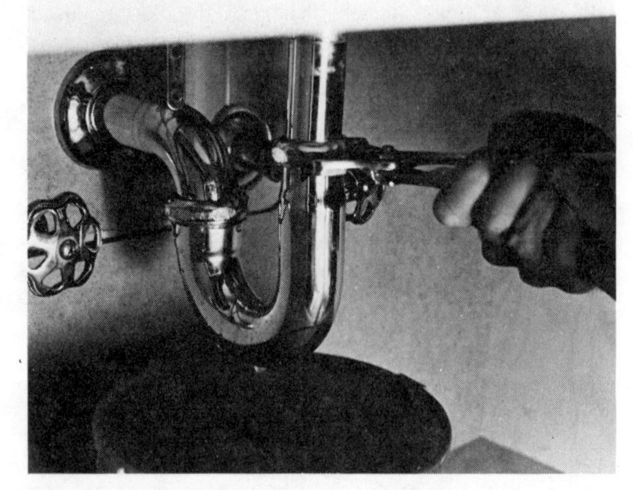

Remove J-bend *by loosening slip nuts at both ends. Bucket underneath catches spillage from the drainpipe.*

Use a coat hanger to clean out the trap. If the obstruction is farther down in the system, use the auger, inserting it into the drainpipe at the wall, or try the other methods shown at right. When the trap is clear, replace all the washers and gaskets and reinstall the trap.

Other methods. You can use any of the methods pictured below and at right to attack an obstruction.

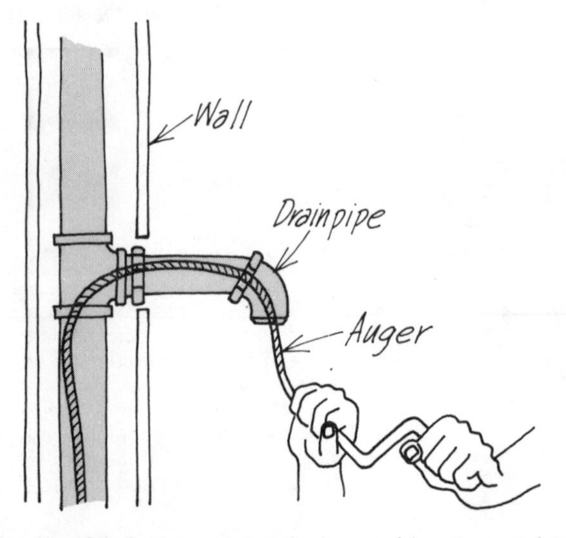

For deep blockages, *remove the trap and insert auger into the drainpipe where it passes through the wall.*

Caution: Before inserting a hose into a drain pipe, turn on the water and keep it running until you remove the hose in order to prevent any backflow of sewage into the water system through the hose. (This isn't necessary if your supply system has a vacuum breaker or anti-siphon valve attached to it.) Remember that the hose may be contaminated and should be cleaned thoroughly after use.

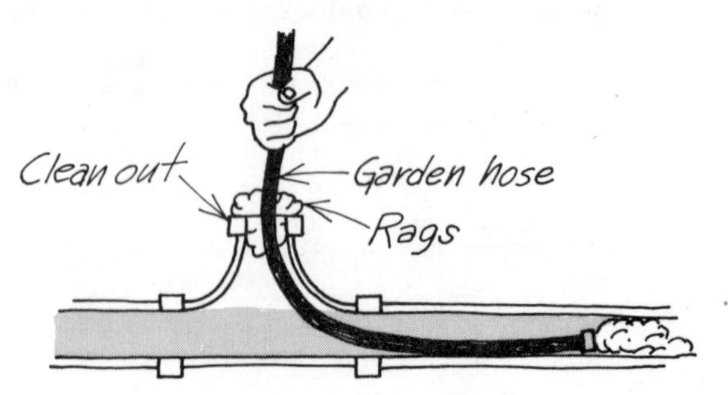

Insert garden hose *into the pipe until it reaches the obstruction, and then turn water on full blast. (If the obstruction is in tight, the water might not dislodge it but instead come shooting back out the open end of the pipe.)*

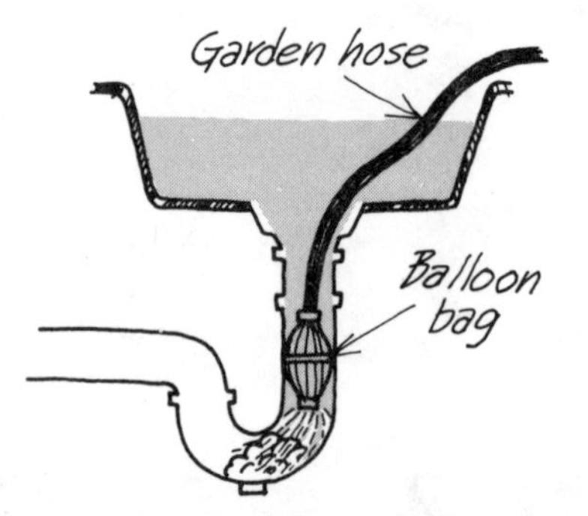

Balloon bag *attaches to a hose and is inserted into the drain. When the water is turned on, the bag expands to form a seal so water can't escape back through the open end of the drain. The resulting build up of water pressure acts to break up most common blockages. Make sure the bag is securely connected to the hose. If you insert the hose in a cleanout plug, be sure it's past any waste line leading from the fixture. If it's not, the water will flow into that line and overflow the fixture.*

If other methods don't work, *try inserting hose or long auger into drainpipe through roof vent. Turn on hose full blast.*

Clogged shower and tub drains

To unblock drains in showers and tubs, use the same procedures described on pages 21-22 for clogged sink drains.

A trap for a tub isn't necessarily the same as a sink trap. Instead, a tub may have a drum trap, and if the obstruction is there you'll have to remove the trap's cover to get to it.

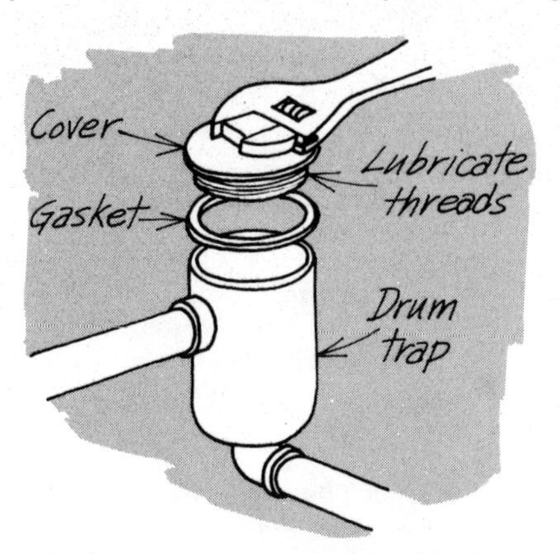

Drum traps *for bathtubs are common in older houses. Solids settle in bottom, water drains out through top pipe.*

To loosen stubborn trap cover, *cut notches with cold chisel, use the hammer with a punch or a drift pin to force cover (counterclockwise). Clean out the trap and replace the cover.*

Clogged house drain or sewer

If more than one drain is blocked or sluggish, the obstruction is either in the main house drain, which collects waste from all fixtures, or in the house sewer, the underground drain that goes from your house to the sewer main under the street or to your septic tank. There's a cleanout plug for each of these. Find it, put a pail underneath (you may need several), have some papers and rags on hand, and open the plug slowly so that you can control the flow of waste. Use any of the methods on the opposite page for clearing it.

Main Drainpipe. For clearing a blockage in the house drain, use any one of the methods shown on the opposite page—augering, garden hose, or balloon bag. If the obstruction is close to the cleanout, you may even be able to loosen it with a piece of stiff wire.

House sewer. The cleanout for the house sewer is near the wall of the house, probably outside. For unclogging it, you can use a sewer tape, a strong metal tape rolled on a wheel and fed by hand into the sewer; a sewer rod, which you push into the sewer and force against the blockage; or a power auger. Most equipment rental stores will have the tape, the rod, and the auger.

The most common house sewer blockages occur because tree roots have grown in through the joints in the pipe. Power augers have blades designed specifically for the purpose of cutting roots. An electric motor feeds the auger into the pipe and keeps the auger blade turning.

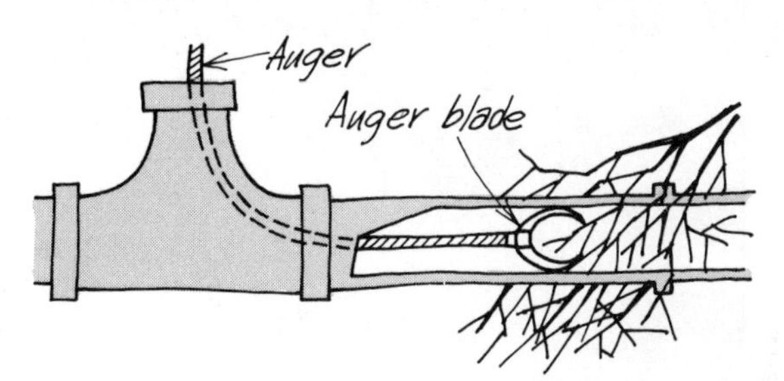

Special auger blade *cuts through tree roots. If roots have damaged pipe, replace that section (see pages 62-77).*

If problems with the house sewer persist, the damage is serious enough to require major repairs on the line. Underground pipe is particularly difficult to work with if you've had no experience. It's best to call a professional plumber.

Clogged floor drain

Floor drains may be clogged with sand, dirt, hair, or lint. To clean them out, remove the strainer and remove obstructions. Sometimes the strainers, especially in basements, are cemented into place. You'll have to chip the cement with a hammer and chisel to free the strainer.

If the trouble is beyond the strainer, try any of the methods already detailed for clearing the blockage.

Unscrew and lift *off floor drain strainer; then reach down into drain to clear out hair, other debris.*

WHEN A DRAIN LEAKS

Leaking around a kitchen sink drain where the strainer and the drainpipe meet is usually caused by the failure of washers in the drain assembly.

The first step in solving the problem is to unscrew the drain tailpiece (the straight piece of pipe that connects with the trap bend) by turning its locknut counterclockwise. To get the pipe out of the way and have space to work in, you'll also have to unscrew it where it connects with the J-bend of the trap. A large metal nut holds the drain in place. Use a spud wrench or a hammer and a screwdriver, as shown, to remove it.

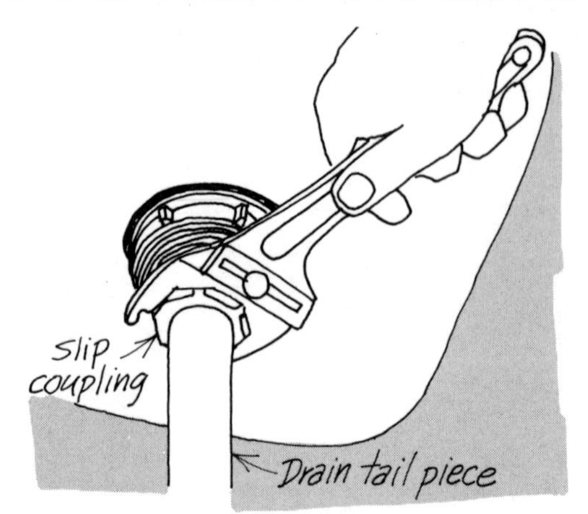

Step 1. *Unscrew locknut to remove drain tailpiece.*

From inside the sink, lift the drain out of place and remove the washers (metal or rubber). Get exact replacements for them.

Step 2. *Use adjustable spud wrench to remove large nut that holds drain in place; then lift out drain from top of sink.*

Clean around the outlet hole in the sink and apply a ⅛-inch bead of plumber's putty all around it. Set the drain into place and press down hard. From underneath, replace the washers and screw on the large metal nut as tightly as you can while someone holds the drain from the top to keep it from moving.

Step 3. *Place a ⅛-inch bead of putty all around drain hole, fit in drain, and press down hard, creating seal.*

Before reconnecting the drain tailpiece, replace the rubber washers on each end. Remove excess putty in the sink with a soft cloth.

Though constructed a bit differently, drains in bathroom sinks are installed in essentially the same way. Leaking here also means failure of a washer. Steps for removing drain, changing washers, and reinstalling are basically the same as for a kitchen sink.

A laundry tub drain may leak because the gasket has failed or because the screws that hold the gasket in place have been loosened. The assembly, shown below, is mechanically simple. However, rust frequently makes disassembling this type of drain difficult. If you have a leak and rusted screws, treat the screws with penetrating oil. If possible, give the solvent a full day to work—but no less than an hour. Take the old gasket to a plumbing supply store to get the correct replacement.

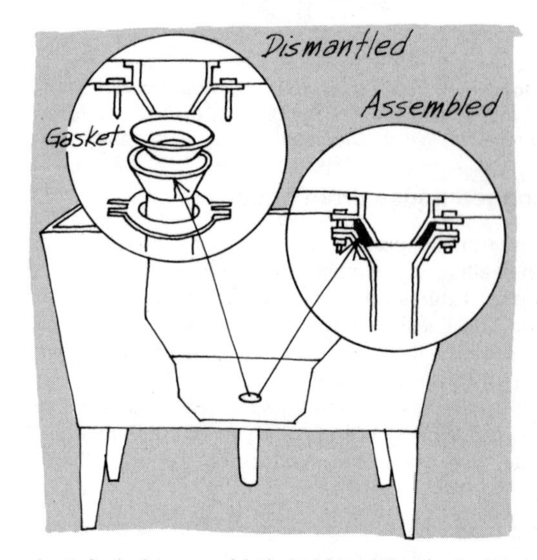

Laundry tub *drain assembly is held together by two screws. Loosen screws and disassemble drain to replace gasket.*

This section gives you solutions to common problems with pipes: leaks, noise, high or low pressure, and freezing. If you are replacing old pipes or adding new ones, see page 62-77 for information on pipefitting and taking pipes apart.

Leaky pipe

A higher-than-normal water bill might give you the first indication of a leaking pipe. Or you might hear the sound of running water even when all fixtures are turned off. When you suspect a leak, check first at the fixtures to make certain all the faucets are tightly closed. Then go to the water meter. If the dial moves, you're losing water somewhere in the system.

Finding the location of a leaking pipe isn't always easy. The sound of running water helps; if you hear it, follow it to its source.

If water stains or drips from a ceiling, the leak will probably be directly above. Occasionally, though, water may travel along a joist and then stain or drip at a point some distance from the leak. If water stains a wall, it means a leak in a vertical section of pipe. Any stain is likely to be below the actual location of the leak, and you'll probably need to remove an entire vertical section of the wall to find it.

Without the sound of running water or without drips or stains as tangible symptoms, leaks are more difficult to find. Use a flashlight and start by checking all the pipes under the house in the crawl space or basement. It's quite likely you'll find the leak there, since leaks other places in the house will usually make themselves apparent with water stains or dripping.

Patching a leak in a pipe is a simple task if the leak is small. The ultimate solution is to replace the pipe, but here are temporary solutions until you have time for the replacement job (see page 76 for instructions on taking pipe apart).

The methods shown below for patching pipes are all effective for small leaks. The clamps should stop most leaks for several months. All clamps should be used with a solid rubber blanket. Buy a sheet of rubber, as well as some clamps, at a hardware store and keep them on hand for this purpose.

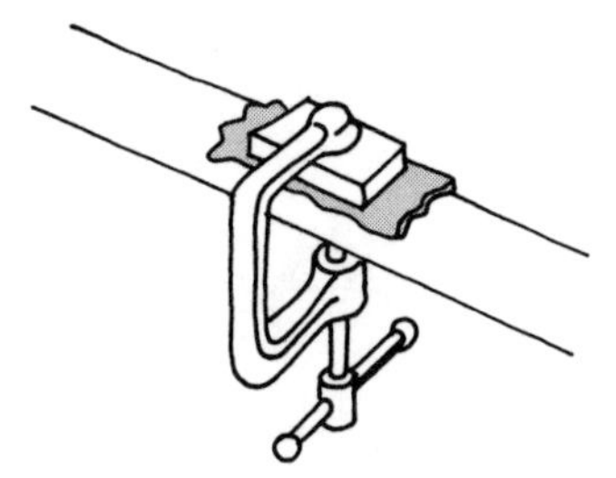

C-clamp *and a small block of wood will stop leak when nothing else is at hand. Use rubber here, too.*

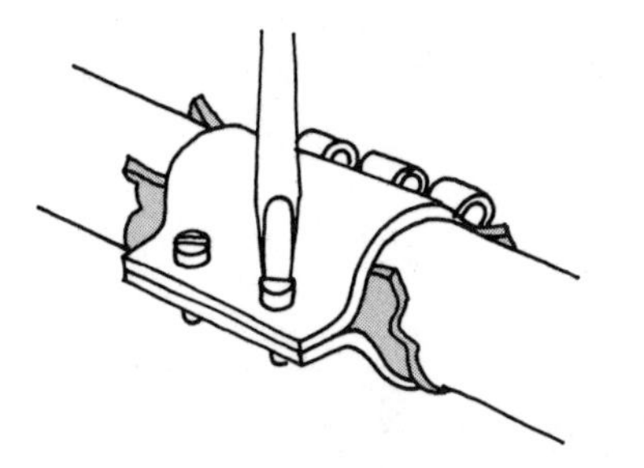

Sleeve clamp *that fits pipe diameter exactly works best. Use rubber blanket over leak, screw clamp down tight.*

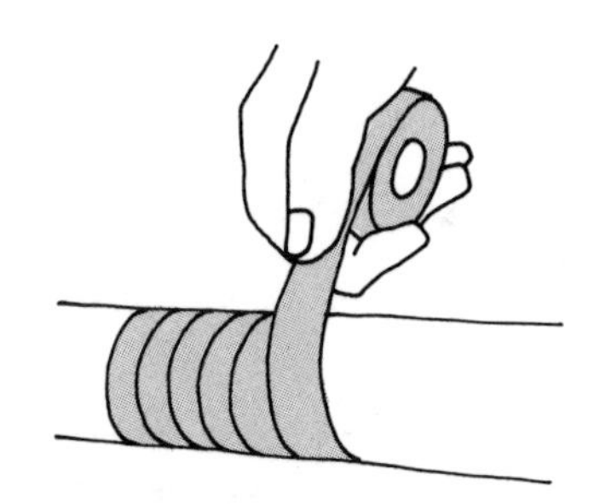

Wrap three layers *of plastic electrician's tape 3 inches on either side of pinhole leak. Overlap each turn by half.*

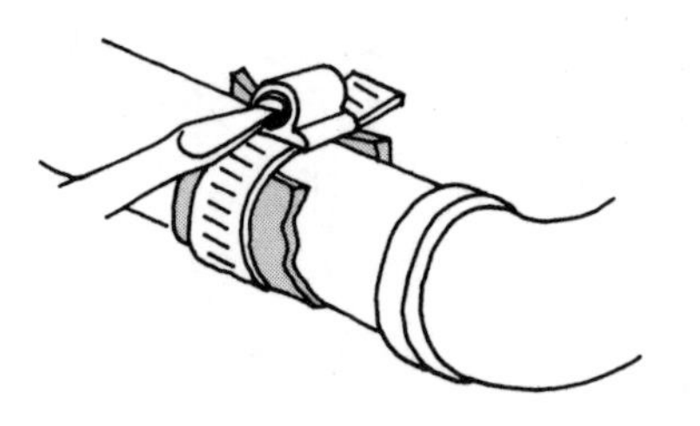

Hose clamp *(size 16 or 12) stops pinhole leak on any size pipe. Be sure to use with a rubber blanket.*

Epoxy putty *will stop leaks around joints where clamps won't work. Pipe must be dry for putty to adhere to it.*

Leaky valve

The two kinds of valves you're most likely to find in your plumbing system, apart from those in the fixtures, are globe valves and gate valves (pictured below). The globe valve has a washer at the end of the stem; the washer fits into a valve seat to control the water flow. The gate valve has a tapered part at the end of the stem to control the water.

These valves may leak around their stems. In that case, close the valve tight and unscrew the packing nut. If there's a packing gland beneath it, lift it up. Wrap the stem with a few turns of strand graphite packing (available at hardware and plumbing supply stores), replace the gland, and tightly screw down the packing nut.

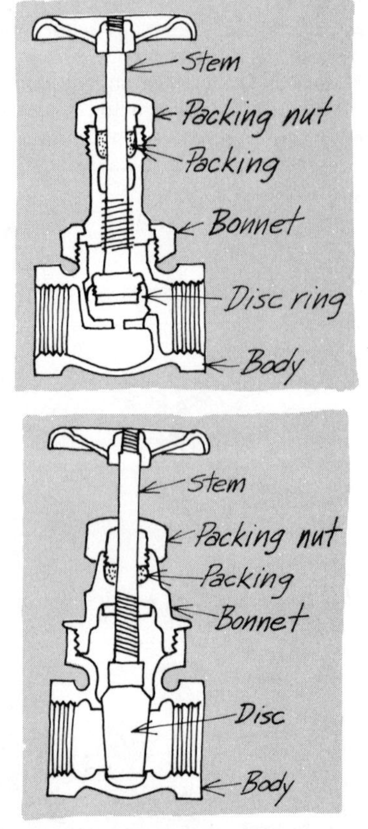

Globe valve *(above) stops water flow with washer. Gate valve (below) controls water with disc at end of stem.*

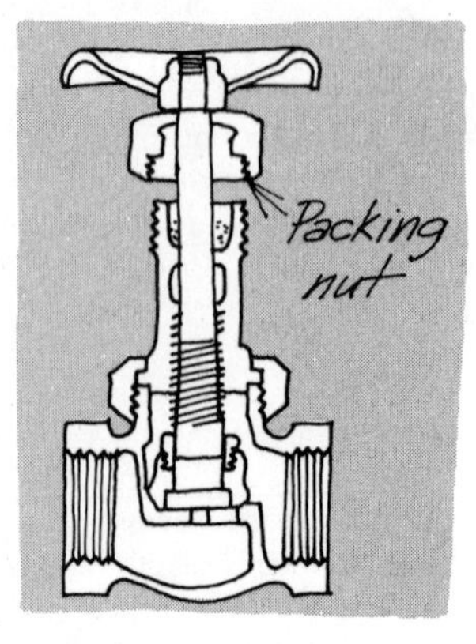

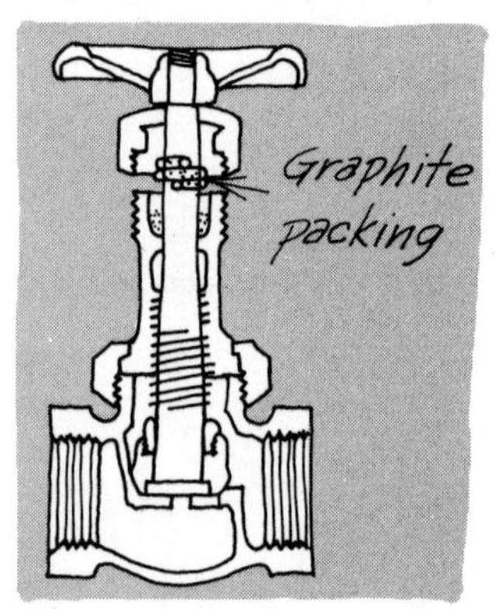

Repack *a leaking valve without removing the handle.*

Patch that cracked sink or tub

Laundry tubs made of cast cement become porous and fragile with age. If a tub begins to seep water through decomposing cement, there's nothing to do but replace it. Neither paint nor any other sealing compound can bond to the soap-saturated surface.

Cracks, though, can be patched to prolong the life of the tub. Regular cement, mixed in finish proportions (1 part cement, 3 parts sand or a bit less) is adequate but not as good as the special cements.

One of these cements uses a liquid latex binding agent in place of water. Another has a plastic vinyl binding agent mixed in with the cement powder; this type uses water.

Porcelain sinks that are cracked or chipped can be temporarily repaired with one of the products available for this purpose. They blend in fairly well but need to be renewed regularly.

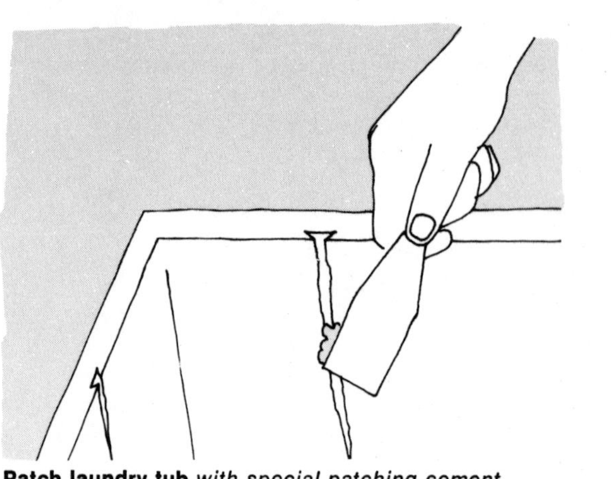

Patch laundry tub *with special patching cement.*

Noisy pipes

If your pipes are noisy, first check the way they're anchored —it's the easiest problem to solve. Pipes not anchored securely to the house framework may shudder when a valve is closed. U-brackets will secure the pipes. A rubber or felt blanket in the bracket adds a cushion against banging and rattling.

To prevent banging noise, *pipe is braced by wooden block and held securely to it by metal U-bracket nailed in place.*

If the pipes are anchored securely, the probable cause of noise is a condition called "water hammer" (the sound of pipes shuddering and banging) which results from a loss of air cushion in the system or no air cushion's having been designed into the system in the first place.

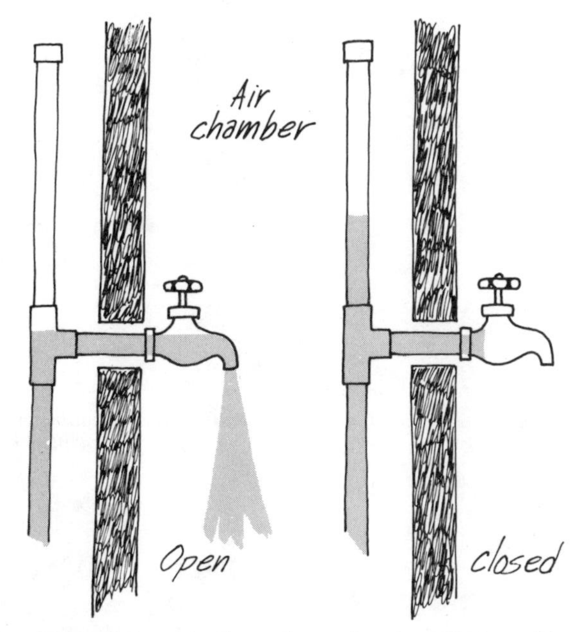

When faucet *is turned off, moving water doesn't stop with a bang but rises in the air chamber, preventing "hammer."*

Restoring air cushion. Most water systems have short sections of vertical pipe rising above each faucet. These sections hold air that cushions the shock when flowing water is stopped by a closing valve—the moving water rises in the pipe instead of banging to an abrupt stop. Sometimes these sections get completely filled with water and lose their effectiveness as cushions. To restore air, take these steps:

1. Check toilet tank to be sure it is full and then shut off valve just below the tank.
2. Close the main shutoff valve for the house water supply.
3. Open the highest and lowest faucets in the house to drain all water from pipes.
4. Close the two faucets; reopen the house valve and the valve below the toilet tank. Normal water flow will reestablish itself for each faucet when you turn it on. (You can expect a few grumbles from the pipes before the first water arrives.)

Installing new air cushions. If the plumbing system in your house was designed without air cushions or if some individual fixtures have none, you can purchase air cushions at a plumbing supply store and install them yourself. You can buy individual ones—known as air chambers, water hammer arresters, or air cushions—or purchase a master unit for the entire house.

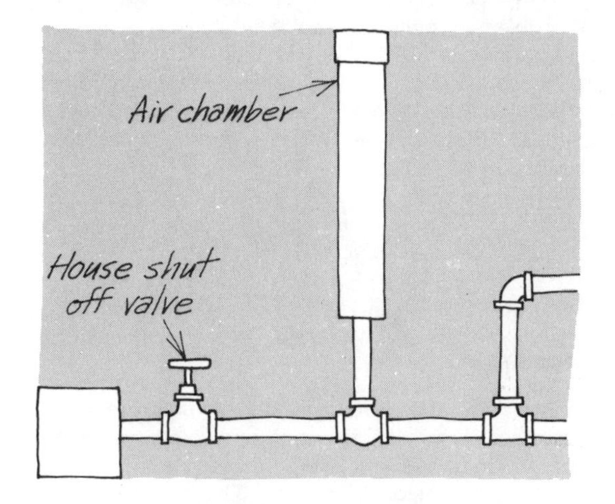

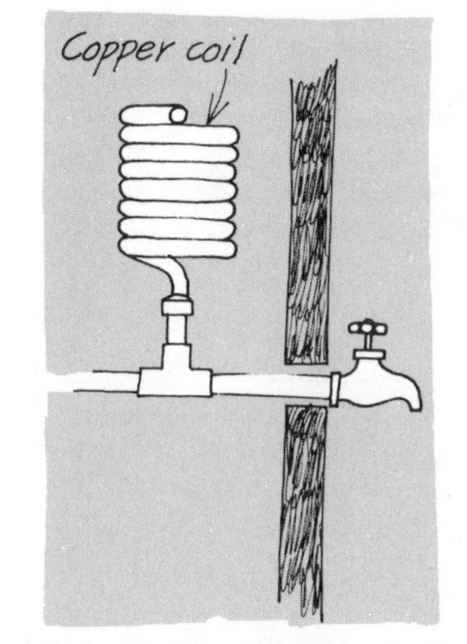

Air chambers *that serve the whole house, instead of individual fixtures, can be straight or coiled.*

Pressure problems

Most appliances and valves that use water are engineered to take 50 to 60 psi. Mains deliver water at pressures as high as 150 psi and as low as 10 psi. Too much pressure is much simpler to cure than too little pressure.

Low pressure. The symptom of low pressure is a very thin trickle of water from faucets throughout the house. Chronic low pressure is usually found in homes on hills above reservoir level or in old homes whose pipes are badly clogged by scale and rust. In many communities periodic low pressure may occur during peak service hours through no fault of the home's location or plumbing.

Whatever the cause, cures are either very expensive or mechanically infeasible. They range from complete replacement of the plumbing to building your own reservoir in a tower tank above the house.

Still, you can make modest improvements. Systems with the beginnings of clogged pipes can regain some lost pressure if you flush the system. To do this, take these steps:

1. Remove and clean aerators on faucets.
2. Close the valve that controls the line you intend to clean. This gate valve may be the inlet valve on the water heater, or it may be the main house valve.
3. Open wide the faucet at the point farthest from the valve, and open a second faucet nearer the valve.
4. With a rag, plug (but don't shut off) the faucet nearer the valve.
5. Reopen the gate valve and let water run full force through the farther faucet for as long as sediment continues to appear—probably only a few minutes. Close faucets, remove rag and replace aerators.

Before you decide to replace all the pipes in your house because of low pressure, try replacing the one section that leads from the main to your house. If it's a ¾-inch pipe, replace it with a 1-inch pipe. The larger size won't increase the pressure, but the increased volume of water will compensate. You can also ask your utility company to install a larger meter.

High Pressure. The symptoms of high pressure are loud clangs when dishwashers shut off or wild sprays when faucets are first turned on. High pressure usually occurs in houses on lower slopes of steep hills or in subdivisions where high pressure is maintained as a matter of fire protection. Above-normal pressure can be cured easily and inexpensively by the installation of a pressure-reducing valve (see illustration below).

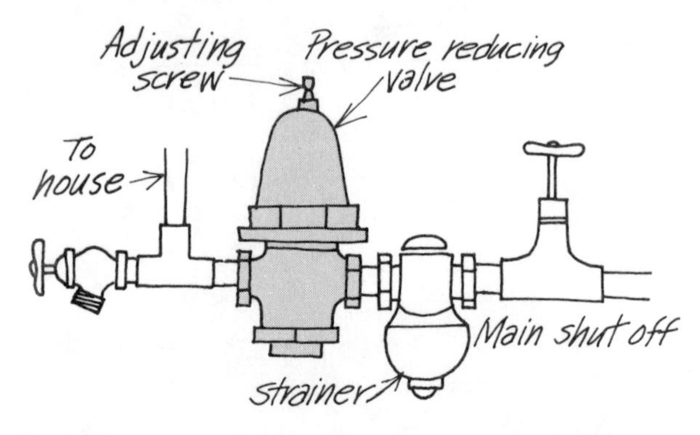

Install pressure-reducing valve *between meter and main shutoff valve. Screw at top of valve regulates pressure.*

Thawing frozen pipes

To prevent pipes from freezing, see page 29. If pipes do freeze, first turn off the water supply and then open the outlet nearest the frozen section. Use one of the following ways to thaw the pipe, working toward the outlet so the melt will have a place to run off.

Heat pipe *with torch. Be very careful not to get pipe hotter than you can touch or buildup of steam may cause an explosion. Protect surrounding wood with asbestos panel.*

Wrap the frozen section *in rags and pour boiling water on them. Rags hold water's heat where it's needed.*

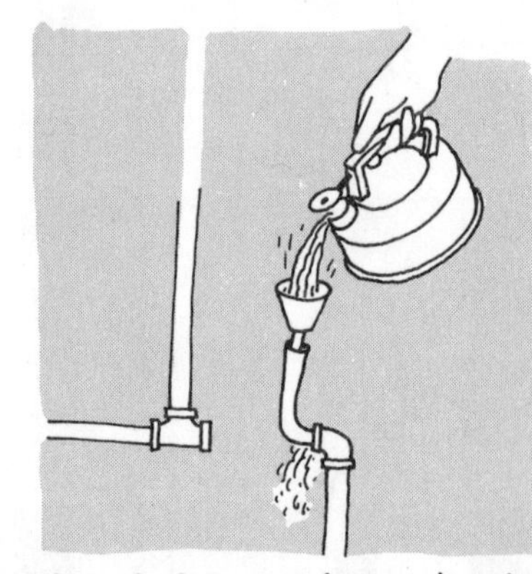

Remove section of pipe *near freeze, insert rubber hose and funnel, pour in hot water. Melt will run out pipe end.*

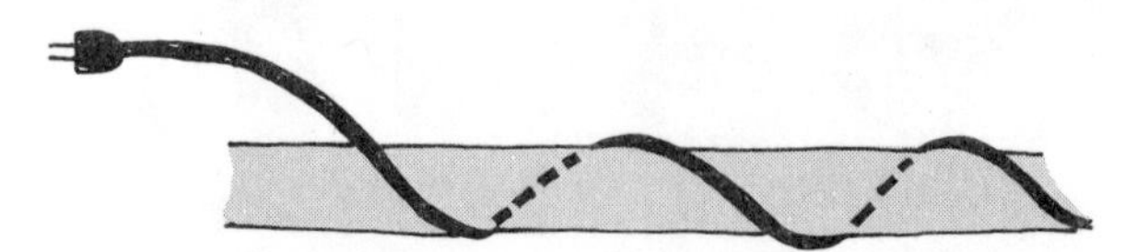

Electric heating cable *around pipes will melt ice in them. It's also good prevention; plug in on cold nights.*

Should pipes be insulated?

The most common reason for insulating pipes is to protect those that are likely to freeze. Consider insulating any pipes on poorly insulated outside walls or all pipes if you leave a home in a cold-winter area unheated while water remains in the system.

In addition to keeping pipes from freezing, insulation on your pipes minimizes heat loss from the hot-water system. Metal pipes are especially good conductors of heat and can conduct a significant amount of heat right out of your hot water.

Insulation also prevents cold-water pipes from sweating. When water in pipes cools and the cold metal meets the warm surrounding air, condensation occurs. With pipes installed in walls and above ceilings, constant sweating can cause rotting in the surrounding wood. (Plastic pipe, because it's self-insulated, presents much less problem with sweating.)

Insulation comes in several forms: asbestos tape, self-sticking tape, liquid that can be applied with a brush like paint, and various kinds of pipe jackets.

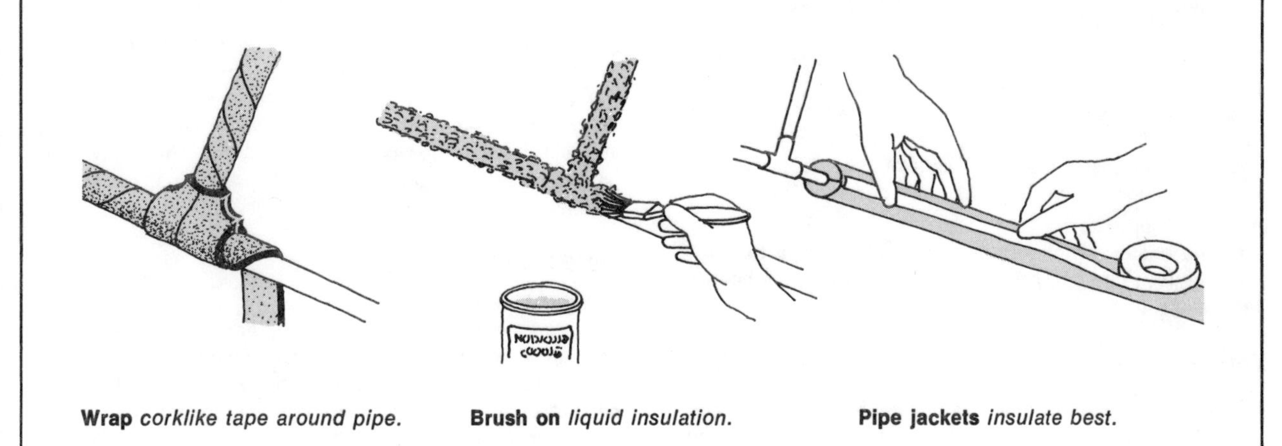

Wrap *corklike tape around pipe.* **Brush on** *liquid insulation.* **Pipe jackets** *insulate best.*

IF THE GARBAGE DISPOSER STOPS

Some disposers are equipped with an automatic shutoff that prevents the motor from operating when overloading occurs. This safety feature protects both the house wiring and the appliance.

With another type of cut-out switch, the grinding mechanism stops when overloaded, but the motor keeps running. In either case, you must turn the switch off, wait 3 to 5 minutes for the motor to cool, and then press the reset button at the bottom of the motor.

Overloading is generally caused by forcing waste material into the unit, or by putting in material it cannot digest. (Styrene foam, plastic, string, aluminum foil, wire, and other metal will jam any disposer. Also avoid glass, paper, and fibrous vegetable matter.)

The common method of unjamming a disposer is by applying pressure against one of the impellers (cutting blades) on the turntable. Follow these steps:

1. Turn off the switch.
2. Place a 1 by 2-inch board (or broom handle) so it rests against one of the impellers.
3. Using the wood piece as a lever, work the turntable back and forth. Repeat this motion until the jam is dislodged.
4. Push the reset button to restart the motor.

Aside from avoiding materials that jam disposers, these two additional points are worth observing:

- Always use cold water to flush the disposer, because cold congeals grease and keeps it from coating drain pipes.
- Don't use caustic drain cleaners in disposers. The cleaners damage seals and grinding chambers. There never should be a need for cleaners if enough cold water is used to wash away material as it's ground up.

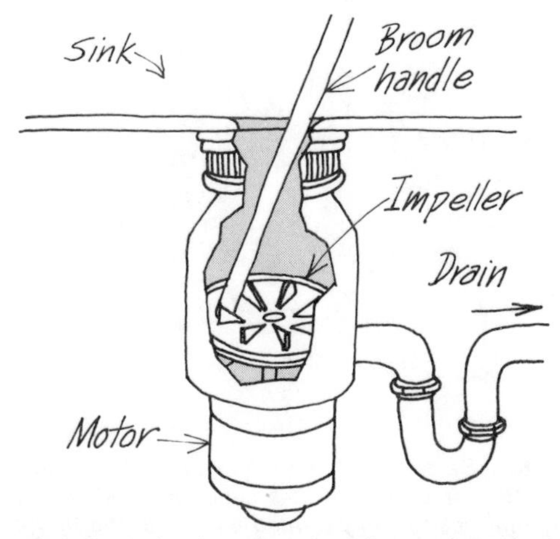

Broom handle *lodged against impeller and worked back and forth should dislodge stuck material.*

TROUBLESHOOTING A WATER HEATER

If you discover water leaking from your water heater, check to see if it is coming from the pressure relief valve or the drain valve (see below). Should you find instead that the water is coming from a hole in the tank, you will probably have to replace the entire tank. In that case, consider whether the present tank is large enough for your needs. See page 44 for information on replacing a water heater.

Here are steps to take to troubleshoot a water heater:

No hot water

In a gas-fired heater:

1. Pilot light is extinguished. Relight according to instructions on tank.
2. Gas supply valve on supply pipe has been inadvertently closed (turned so handle is at right angles to direction of pipe). Open it so that the handle is parallel to the pipe. Then relight the pilot according to instructions on the tank.
3. The thermostat has been shut off. If the thermostat switch is exposed, turn it to the "on" position.

In an electric heater:

1. The heater switch has been shut off. Turn it to "on."
2. The fuse has blown in the circuit supplying the heater. Restore the circuit.

Not enough hot water or water not hot enough

Whether gas or electric, a heater may not hold enough water to meet your needs. Small heaters need as much as half an hour to recover from a full bath or long washing machine cycle. Before you install a larger heater, though, check the following possibilities:

In a gas-fired heater:

1. Thermostat dial is set too low. If thermostat is in view, adjust upward. Normal setting is 140°. If you can't find the thermostat, it is probably buried in the insulating layer around the tank and adjusting will require a visit by the utility company.
2. Valve on supply line is partially closed. Open full—so the handle parallels the pipe exactly.

In an electric heater: Any difficulty with a thermostat will require a visit by your utility company. The thermostat must be buried in the insulation to work properly, so it is out of reach. Disturbing the insulation is almost certain to make a thermostat function less accurately than before.

If a heating element in an electric water heater has failed, you will need professional repair service. The utility company service department does not make repairs, only adjustments.

Water too hot

In a gas-fired heater:

1. Thermostat is set too high. Adjust lower—140° is normal. If steam or boiling water comes from a tap, the thermostat has failed and must be replaced. If this happens, do not shut off the inlet valve. Its being open allows the water to flow back into the main system, keeping pres-

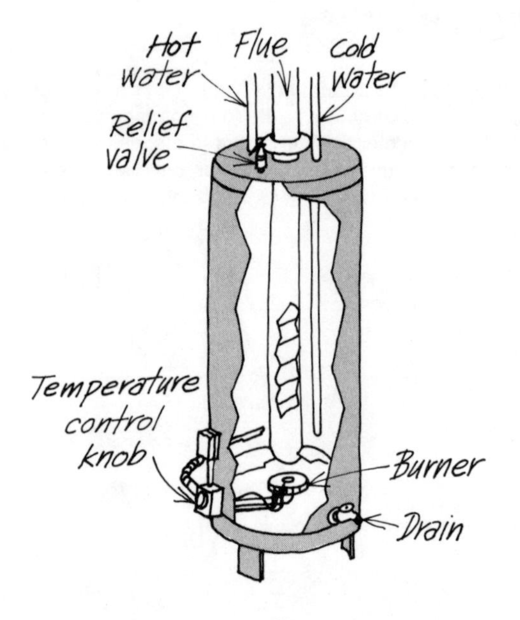

Gas heater. *Be sure you know how to light the pilot light. The heater should come with a handbook and should also have lighting instructions printed on it near the thermostat. It's also very important to have the flue connected properly so combustion residue will escape harmlessly.*

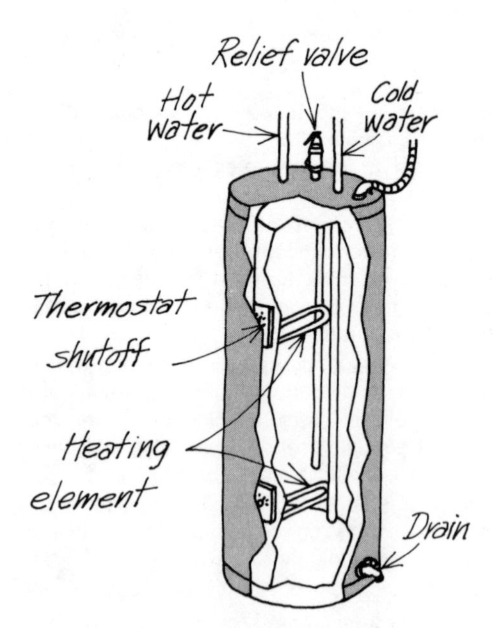

Electric heater. *The thermostat is usually concealed behind a metal door on the side of the tank. If you need to disconnect the wiring to the heater, remember first to shut off the power to the heater at the fuse box or circuit breaker before working with the wires.*

sure below the explosion point. If possible, shut the heater off until you can get professional help. (A heater with a pressure relief valve will not emit steam except through the safety valve.)

In an electric heater:

1. Thermostat is set too high. Often a thermostat is purposely set high to cover peak use periods.

2. Heating element is not grounded. Symptom is excessively hot water only after long periods of the water's not being used. Someone from the utility company will adjust the thermostat; you will need a repair service to ground the heating element.

Noisy water heaters

Water heaters sometimes make very loud rumbling and cracking sounds caused by a buildup of sediment in the tank. As this sediment accumulates, small quantities of water get trapped in the layers. When this trapped water gets very hot, it explodes out of its pocket. The rumbling and cracking are actually a series of tiny steam explosions.

If regular flushing by opening the drain valve every two to six months doesn't solve the problem, you can buy a commercially available cleaning compound that will help dissolve the accumulated scale. Package instructions will give you specific details on how to use the product.

Pressure relief valve

In many communities, laws require that water heaters have a pressure relief valve. Where a home system is "closed" by other plumbing devices so that pressure can't escape back into the mains, such a valve is required.

These valves allow steam to escape safely in case of malfunction in the thermostat. They are invaluable for safety purposes, but they do create something of a nuisance because they tend to leak with every slight surge of pressure.

There is no way to prevent the valve from leaking, since that is its purpose. The only solution is to catch the leaking water. You can do this by attaching a hose to the valve and running it outdoors into a drain. The location of a heater, though, may make this impractical or unsightly.

A second approach to collecting the water is with a waterproof pan under the heater. Ideally, the pan will have a drain hole that lets water escape through the floor to earth or to a drain.

If the pressure relief valve leaks constantly, it is out of adjustment. Have the utility company adjust it to withstand higher pressure.

Repairing a leak in the tank

Repairing a leak in the hot-water storage tank is only a temporary measure. Leaking is due to corrosion, and by the time the tank has corroded through at one point, the entire tank wall is likely to be so corroded that it will be only a matter of time before other leaks appear. The solution is to buy a new heater.

For a temporary solution, repair the leak with a toggle bolt. First drain the tank. Then drill the leaking hole until it is just large enough to accommodate the bolt. Insert the bolt and rubber washer and tighten the nut.

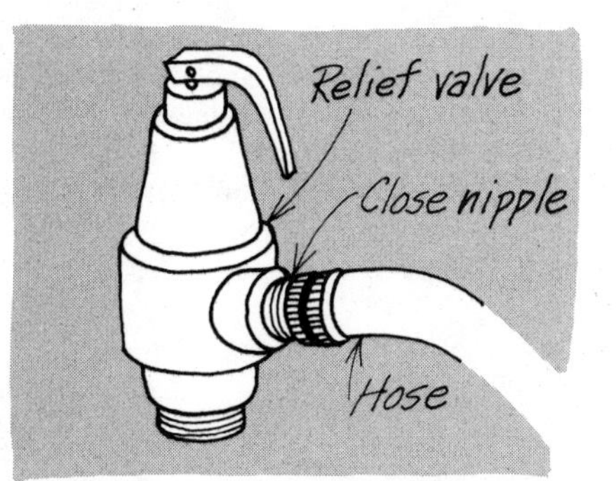

If there's no drain *in floor under heater, attach hose to relief valve and run it to a drain or to outside.*

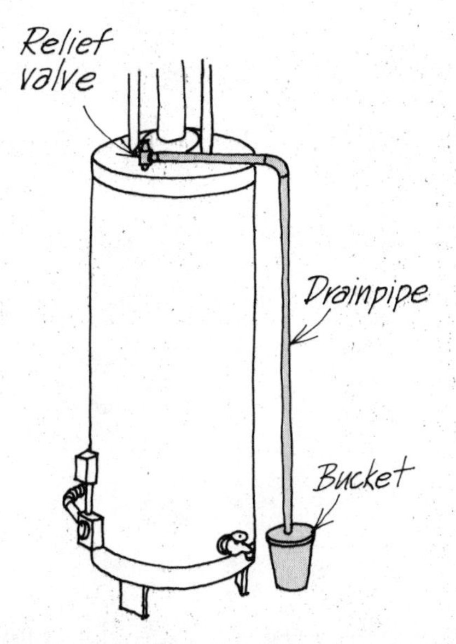

A pan or bucket *under the water heater will catch water from leaking relief valve. Remember to empty it frequently.*

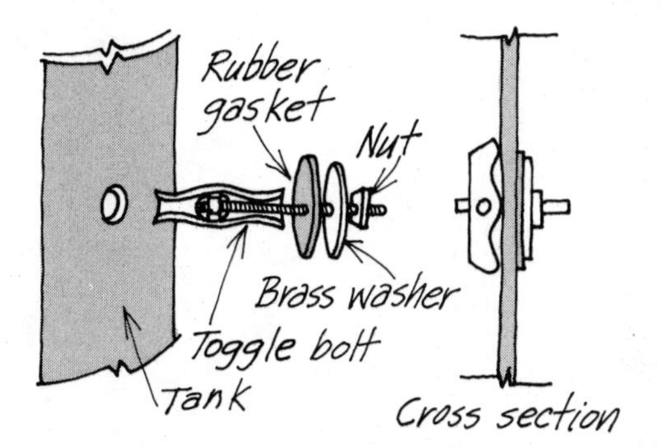

Left. *Drill leaking hole just large enough to accommodate toggle bolt; then insert bolt and tighten it to stop leak.*

Replacing old fixtures

You may have a chronically faulty fixture that has outlived its usefulness. Or you may have old or unsightly fixtures that have outlived their attractiveness. When the time comes to throw anything out—including the kitchen sink—you can save a lot of money by doing it yourself.

Manufacturers of home plumbing fixtures are increasingly conscious of the do-it-yourselfer and many provide clear, specific installation instructions with their products.

When you set out to remove an old fixture and replace it, you're likely to run into some snags if you live in a house more than 15 or 20 years old. The particular style you need may not be manufactured at all any more, and new ones may have slightly different dimensions, requiring some changes in the internal plumbing of the house as well as at the fixture outlet.

When installing new fixtures, be careful not to mar the finish. Use electrician's tape on the jaws of wrenches when working with chrome-plated nuts. (A wrench with smooth jaws is less likely to scratch the chrome than one with toothed jaws. Still, the tape is a good precaution.) When working with enameled or porcelain fixtures, cover the floor or other surface of your work with papers or rags. Tape paper onto the fixture where any part of its finish might be marred by rubbing against another surface.

SINK, TUB, AND SHOWER FAUCETS

If you're replacing a sink, tub, or shower, you will probably want to replace the old faucets as well, but the reverse isn't necessarily true—because faucet assemblies are detachable from the fixture, you can add a new faucet without replacing the fixture.

Several kinds of faucets are available for home use. Single-handle faucets that combine the hot and cold water with one control are popular. New single-handle faucets for kitchen and bathroom sinks will usually fit onto the present plumbing for single or double faucets.

The compression faucet has one handle for hot water and another one for cold. Another type of two-handle faucet replaces the old washer and valve seat of a compression faucet with internal rotating discs.

For detailed information on how all these faucets function, see pages 8-12.

Pick a new faucet

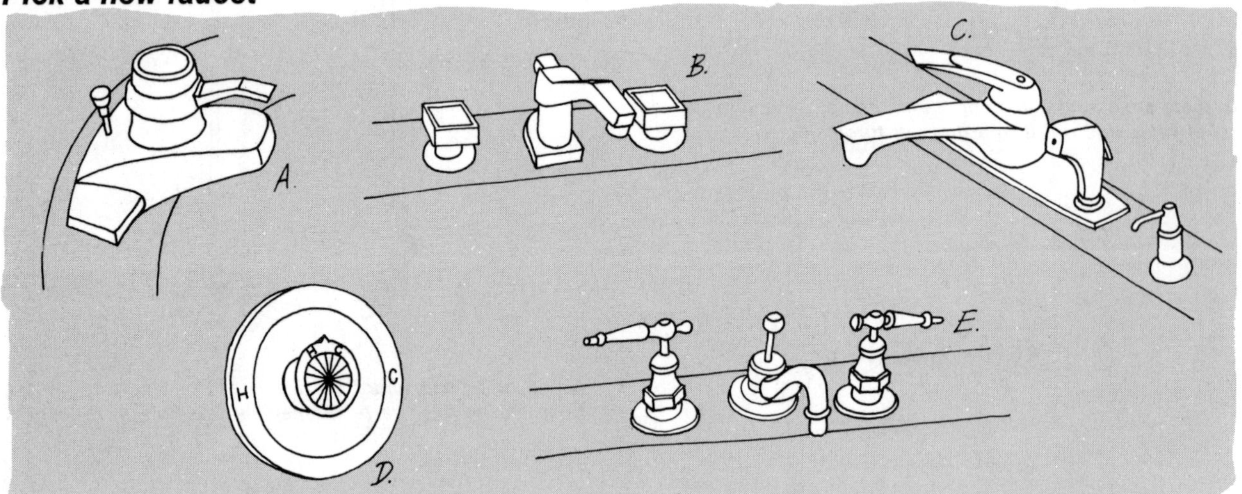

A. *Lavatory faucet has single lever, no washers.* **B.** *Handle inserts on modern faucet are changeable.* **C.** *Spray hose, soap/lotion dispenser are handy to kitchen faucet.* **D.** *Single-handle tub and shower faucet.* **E.** *Antique style faucet.*

Removing a deck-mounted faucet

Deck-mounted faucets are those that fit right onto the sink through precut holes. Most kitchen and bathroom sink faucet assemblies are deck mounted.

The first step is to turn off the water. If there are shutoff valves at the fixture, use them. If not, turn off the entire house water supply at the gate valve where the water supply enters the house.

Lengths of flexible pipe attach to the fixture shutoff valves or directly to the supply pipe stub-outs at one end and to the faucet body at the other. These pipes are held in place by nuts at both ends. Loosen these nuts and remove the pipes.

Since the clearances for working under a sink are often quite limited, getting underneath to tighten or loosen nuts can be awkward. A special tool called a "basin wrench" (pictured below right) is made just for the purpose of working within these tight clearances. Plumbing supply stores have them at low cost. The head of the wrench fits over the nut underneath the sink. The handle extends down at a right angle to the head and can be maneuvered easily where there's more room. The jaws of the wrench lock onto the nut while the wrench is turning in one direction and disengage when the wrench is turned the other way. To reverse the action, all you do is flip the head of the wrench over to the other side.

After you've removed the lengths of pipe between the faucets and the shutoff valves, then remove the nuts that screw onto the faucet's threaded extensions to hold the faucet tightly in place. If you're working on a bathroom sink and the faucet assembly has a pop-up drain as part of it, disassemble the drain pieces (see below).

When the nuts and washers have been removed (and any attached drain assemblies disconnected) lift the faucet out of the sink.

Deck-mounted faucets

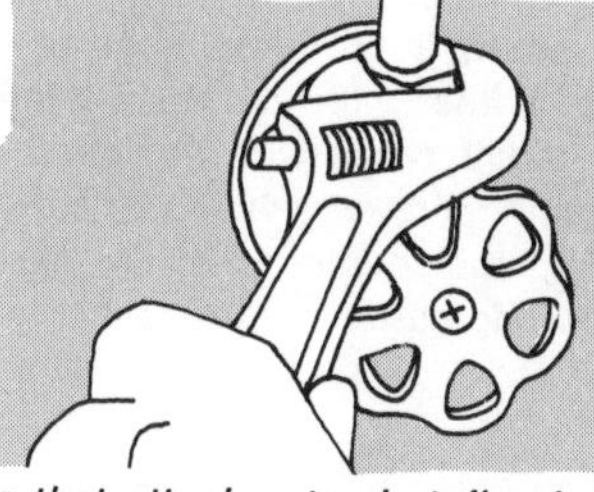

Water flow *from this faucet is controlled by a cartridge. It's single-handled but built to fit onto two-hole deck.*

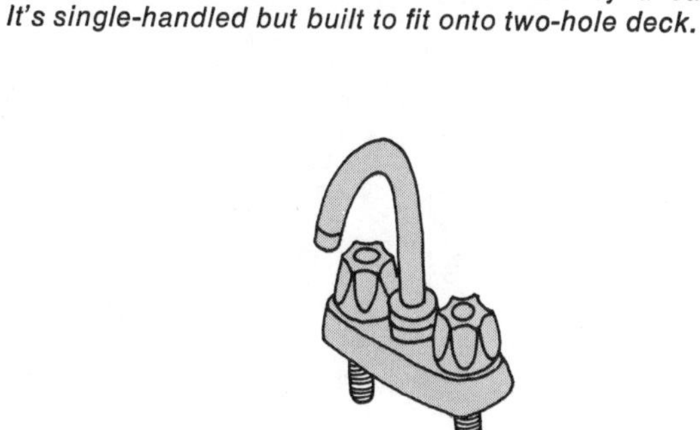

Standard two-handle *faucet with swing spout. Measure distance between holes on sink deck before buying faucet.*

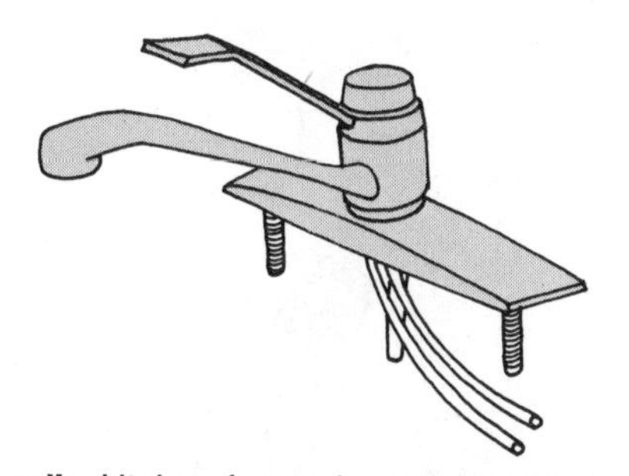

Single-handle *kitchen faucet has copper hookups already attached. They're easily bent to get to supply pipes.*

Removing a deck-mounted faucet

Flexible pipe *that attaches to shutoff valve is held in place by locknuts at both ends. To remove, loosen nuts.*

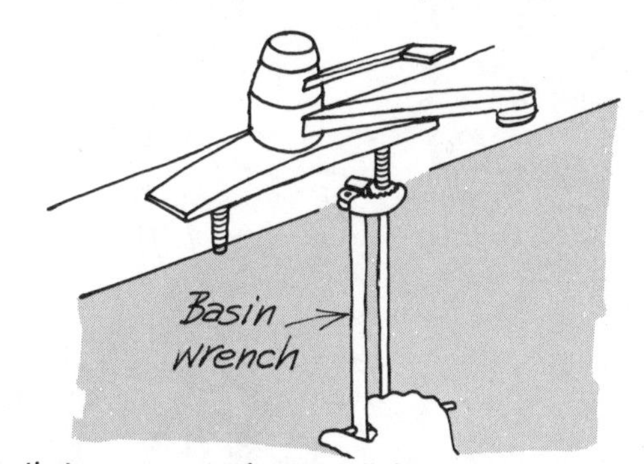

Nuts *that screw onto faucet tailpieces up under sink are hard to get at. Use a basin wrench to loosen them.*

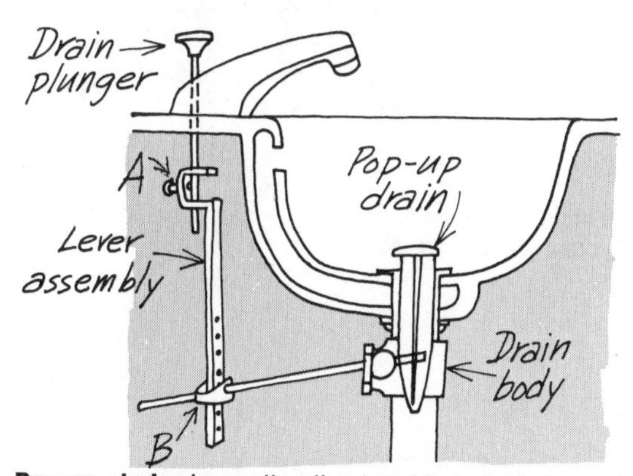

Pop-up drain *is easily disassembled; just unscrew it at points A and B. You must take it apart to remove drain.*

Installing a deck-mounted faucet

Before you buy your new faucet, measure the center-to-center distance between the holes in the sink where the faucet fits. (There may be a third hole in your sink for a spray-hose attachment. This relative distance doesn't matter.) The center-to-center distance will be either 4, 6 or 8 inches. When you buy the new faucet assembly, ask for one with 4, 6, or 8-inch centers.

Some faucets come with the flexible supply pipes already attached. The pipes are usually plain copper, meant to be used where piping is concealed by a cabinet. If they will be exposed, and the existing pipes are damaged or scratched, you might want to buy some new chrome-plated pipes along with the faucet. Since they're flexible, you can bend them to get from the faucet tailpieces to the supply stop valves or stub-outs at the wall.

Clean off the top of the sink with a rag. Put a ⅛-inch bead of plumber's putty all around the bottom of the faucet assembly where it fits flush against the sink top. Fitting the assembly down into the holes, press down.

From under the sink, attach the washers and nuts to the tailpieces and screw them up as tight as you can by hand. Tighten them further with the basin wrench.

If the assembly didn't come with flexible supply pipes attached, attach them first to stop-valves or stub-outs under the sink and then to the tailpieces of the faucet. Use a hacksaw or small tubing cutter to cut them to correct size.

Clean excess putty from around the faucet on the sink top. Before you turn on the water, remove the aerator from the end of the spigot so that any loosened material will flow out freely and wash away. Left in, the aerator is likely to get plugged up. Turn on the water supply and try the new faucet. Leaks mean you haven't tightened the nuts enough or you have forgotten a washer.

Spray hose attachment. The procedure for removing and installing deck-mounted faucets on sinks that have spray-hose attachments is the same as that just described except that you must also disconnect the spray hose. Loosen the locknut that attaches the hose to the faucet body and remove the hose. To get the hose entirely out of the sink, you'll have to remove the coupling at one end or the other. First, unscrew the nozzle; then take off the coupling. It may come unscrewed, but if it's held in place by a metal snap ring, pry it off with a screwdriver. Once the coupling is off, you can pull the hose free of the sink through the guide sleeve in the sink top. The sleeve itself can be removed much as the faucet assembly was—by unscrewing a nut that holds it in place.

If you want a spray hose attachment for your countertop sink and there's no third hole to accommodate it, you can purchase a faucet assembly that has the spray attachment built into it, eliminating the need for an extra hole in the sink or the counter. Or you can drill a hole right next to the sink in the counter itself if the counter is not covered with ceramic tile.

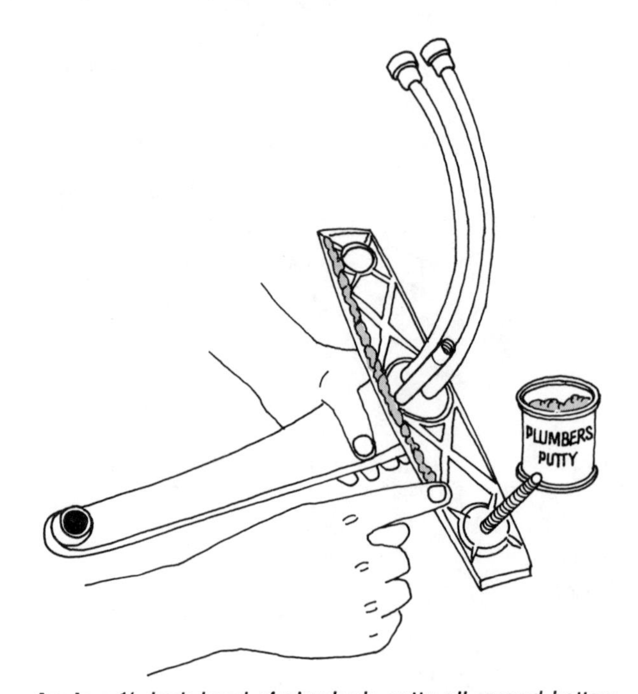

Apply *a ⅛-inch bead of plumber's putty all around bottom of faucet; push the faucet down hard on sink top.*

Spray-hose *attachment for kitchen sink screws onto faucet body underneath; if it's tight quarters, try a basin wrench.*

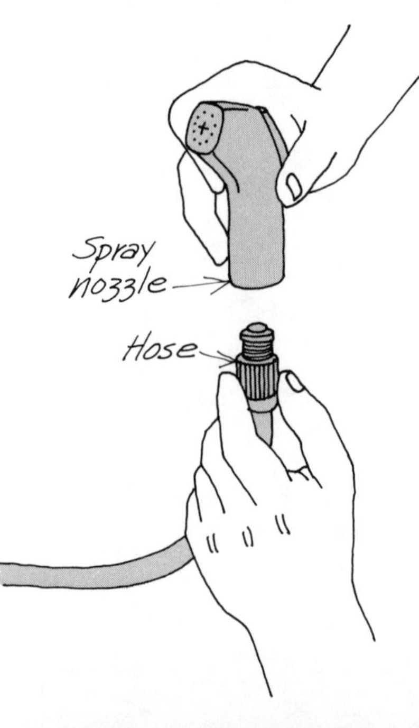

Right. *On some spray assemblies you'll need to unscrew the nozzle and coupling by hand to remove the hose.*

Removing a wall-mounted faucet

Some faucets are mounted directly onto supply pipe stub-outs above the sink instead of onto the counter. These are called "wall-mounted" faucets. Some models screw onto the stub-outs with chrome-plated nuts. Others, such as the single-handle faucet pictured at the bottom of the page, conceal the connections with a decorative escutcheon.

To remove the first type, turn off the water supply to the faucets and unscrew the chrome-plated nuts that hold the faucet in place; then pull the faucet off the wall.

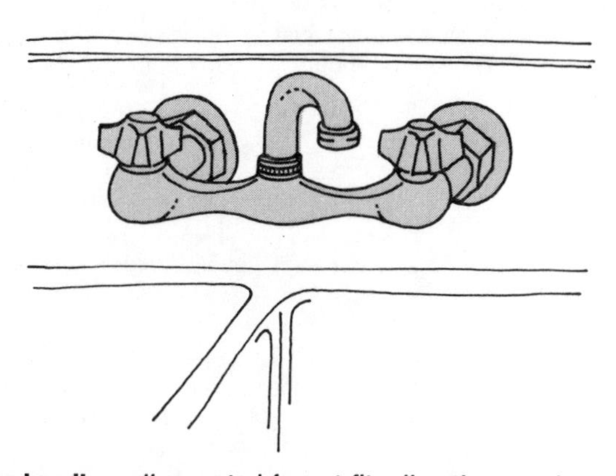

Two-handle *wall-mounted faucet fits directly onto threaded supply outlets on the wall above the kitchen sink.*

To remove the second kind, turn off the water supply to the faucet and then unscrew the spout nut and lift off the spout. You may be able to unscrew it by hand, or you may need a wrench. (Turn it counterclockwise, the same as you do a nut.) Remove the escutcheon next by unscrewing the nut underneath that holds it to the faucet body. Remove the nuts that hold the faucet to the stub-outs and pull the faucet off the wall.

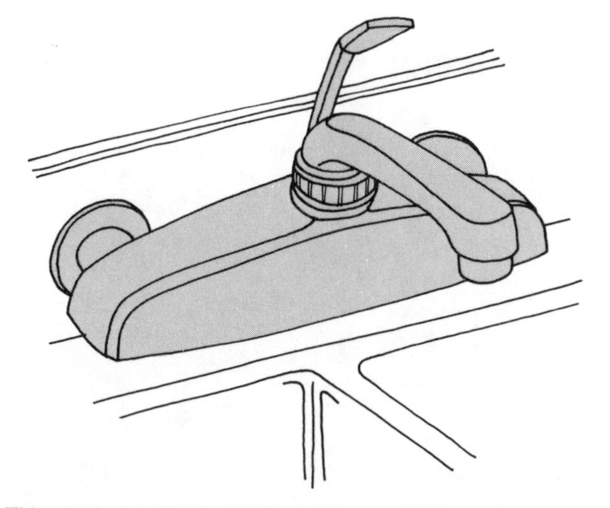

This single-handle *faucet body is covered by a decorative escutcheon held in place by a large nut underneath.*

Installing a wall-mounted faucet

Before buying a new faucet, determine whether the center-to-center distance between supply pipes is 4, 6, or 8 inches. (Some wall-mounted faucets are made to adjust to the space between pipes extending from the wall.) When installing the faucet, be certain to put all gaskets and washers in the places indicated by the manufacturer. Turn on the water supply. Try the faucet and check for leaks.

Replacing a tub or shower faucet

Most tub and shower faucets are wall mounted. They aren't screwed directly into the supply pipes but into the faucet body that's fitted in place on supply pipes behind the wall. If the faucet assembly is an old one, you may have trouble finding new handles to fit onto the body. In that case you're in for more complicated work—you'll have to get behind the wall and replace the faucet body as well.

First turn off the water supply; then unscrew the handles. The handle screws may be under decorative caps (the "H" and "C") that will snap out or unscrew. Remove any decorative escutcheons and unscrew the stem. With tub and shower faucets that are mounted into walls, it's likely that you won't be able to use a standard wrench to remove the stem assembly. In such a case, use a socket wrench (pictured on page 12). Take the old faucet handles with you when you buy the new ones. If you find ones that fit, they'll install easily onto the present body. Be sure to put all washers in place. Screw the bonnets and handles in. Turn on the water and check for leaks.

If you can't find handles to fit or if you want to replace two old faucets with a new single-handle assembly, you'll need to remove the old faucet body and replace it with a new one. To do this you have to get to the pipes behind the wall. Some homes have access to tub and shower plumbing through panels or trap doors in closets or halls that back up against the bathroom. If yours does, your task is simple: just remove the panel. If you find no access panel, you'll have to remove part of the wall.

The diagrams on the next page show how a tub/shower faucet body is mounted onto the water supply pipes. It may be attached with either threaded or soldered connections. If they're threaded, unscrew them. Use one wrench on the

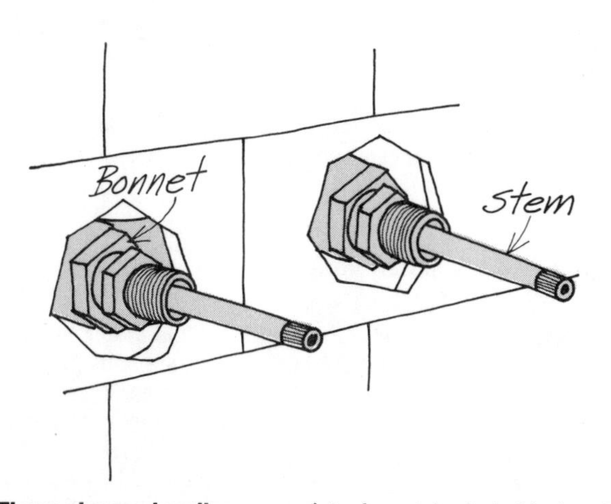

These shower handles *screw into faucet body behind wall. You can unscrew them without getting behind wall.*

supply pipe to hold it steady and one wrench on the locknut, turning it counterclockwise. If the connections are soldered copper ones, unsolder them, using a small torch. (For details on taking apart soldered connections, see page 77.) Take the old faucet body with you when you buy the new one to make certain of correct size.

Installing a tub or shower faucet. Install the new faucet body onto the supply pipes. If the fittings are threaded, apply pipe dope to the threads of the male pipes and screw the locknuts down tight. If the fittings need to be soldered, follow the soldering instructions on pages 68-69. To make certain the pipes can't bang, anchor them firmly with straps before the wall is closed up. When replacing the wall, measure the new assembly carefully to cut holes in the right places for handles.

Adding a shower faucet to a tub

If you want to add a shower over your tub you'll have to cut into the wall and change the plumbing behind the wall. Plumbing supply stores have all the necessary pipes and fittings. You'll need a faucet body that will accept a shower connection and that has a diverter valve to change from tub to shower. Shower pipes, fittings, and head come as a unit.

The work is the same as detailed above, with two additions. First, you'll have to cut a hole in the wall 5 feet above the bottom of the tub for the shower head connection (be sure to take precise measurements before cutting the wall and putting it in place). In addition, you'll have to notch the joists on either side of the shower pipe (see page 53) and nail a 2 by 4 in place just under the outlet to act as a pipe brace. The brace should be flush with the studs.

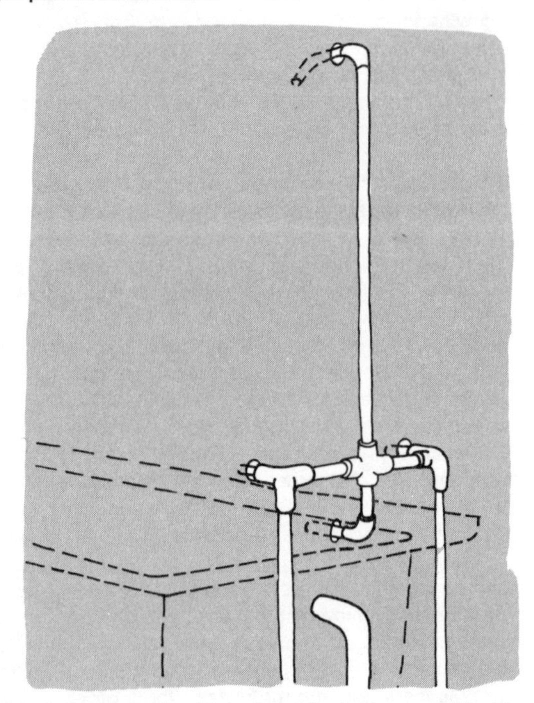

Behind-the-wall plumbing *for tub-shower combination. A diverter valve directs water to tub spigot or shower head.*

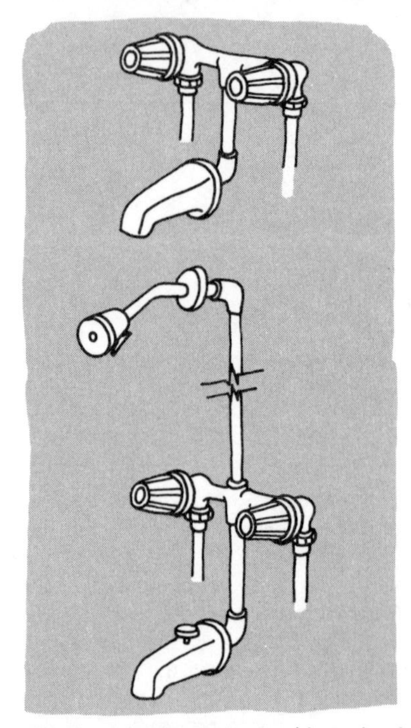

Faucet assembly *for tub/shower (below) is a simple variation of tub-only (above). Diverter in spigot sends water to shower.*

SINKS, TRAPS, AND STRAINERS

In the following section, you'll find instructions for removing and installing kitchen and bathroom sinks and the sink traps and strainers.

Replacing a kitchen sink

To remove a kitchen sink, first turn off the water supply closest to it; then take off the faucets (see previous pages). Remove the trap (see page 40) and the strainer (see page 38). If your counter is tiled, you may have to chip away some tiles to free the sink.

Before you purchase your new sink, be sure to measure the hole in the countertop. Take the measurements with you when you shop for the sink.

Some rimless sinks *are tiled all around their outer edges. To remove the sink, chip away tiles with hammer and chisel.*

Pick a new sink

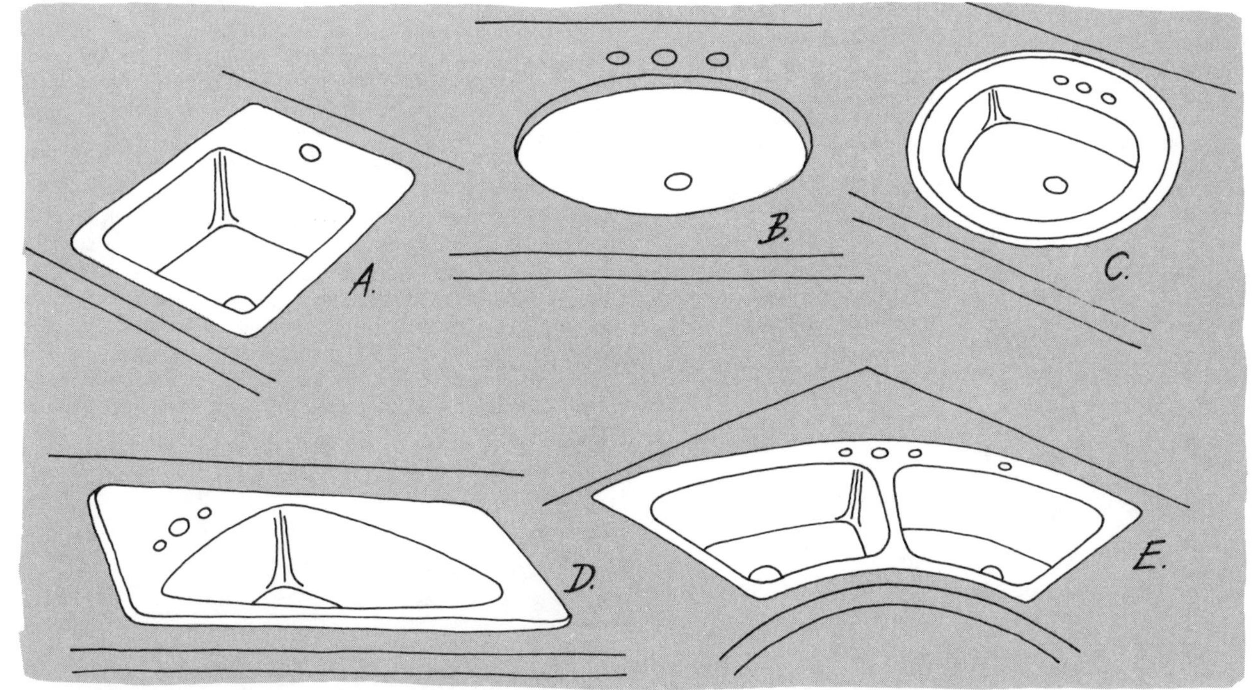

A. Small sink for an "extra." B. Oval lavatory installed under the counter. C. Round, flush-mounted basin rimmed by metal ring. D. Self-rimming basin for narrow counters. E. Double kitchen sink fits into corner.

Stainless steel sinks. These lightweight sinks rest on a lip that fits over the countertop; they require no rim. To remove the sink, unscrew the metal clamps that hold the sink onto the counter after the faucets, trap, and strainer are out.

To install a stainless steel sink, fit the sink into the opening so that it rests on the lip. Attach the metal clamps and tighten screws.

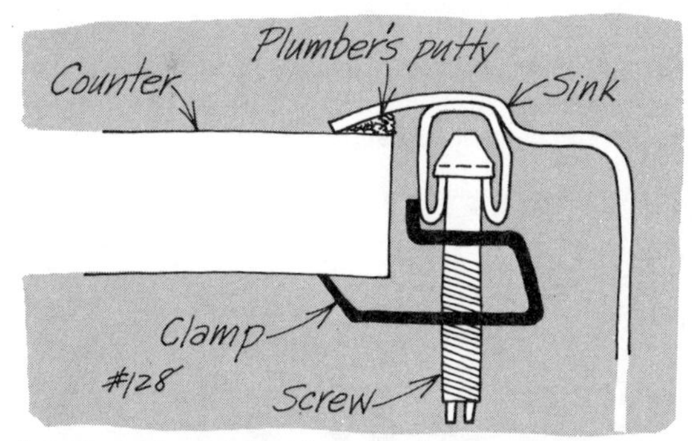

Stainless steel, *self-rimming sink has a clamp to hold it onto the countertop. Tighten screw to secure sink.*

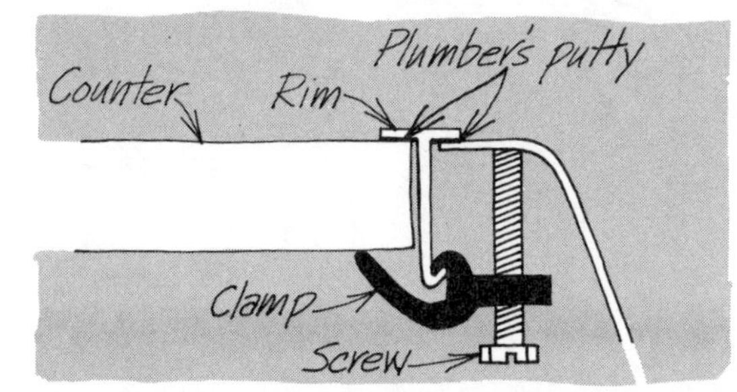

Porcelain-coated steel and cast-iron sinks. Steel sinks are much lighter than cast-iron, but they are removed and installed the same way. You'll need at least two people to lift a cast-iron sink.

To remove a rimmed cast-iron sink, brace the sink from underneath while removing the metal clamps. Then lift the sink and rim out of the counter.

(Continued on next page)

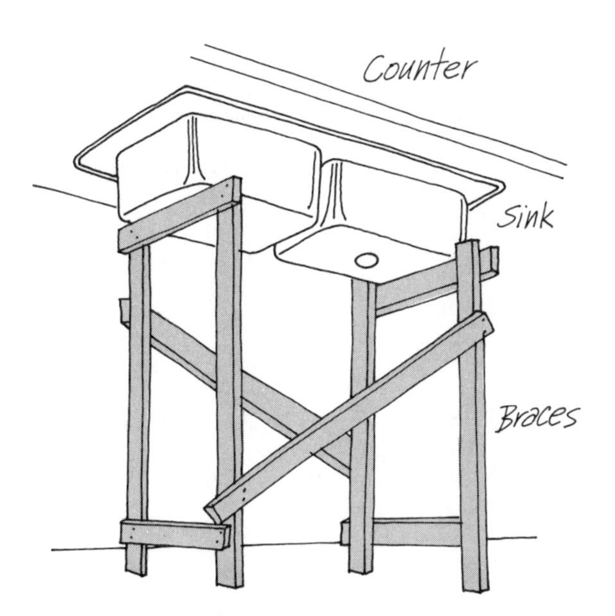

Cast-iron and porcelain coated steel *sinks require rim (left). Brace sink while removing clamps (right).*

. . . Cast-iron sink (cont'd.)

To install this kind of sink, fit the rim onto the sink, using the metal clamps shown below. (If the rim has the special extensions shown, you can fit the rim and sink together as illustrated.) Once the rim is securely in place, lower the sink into the countertop. From underneath, attach the metal clamps shown on the preceding page and tighten the screws. For temporary support you can nail a couple of 2 by 4s under the sink.

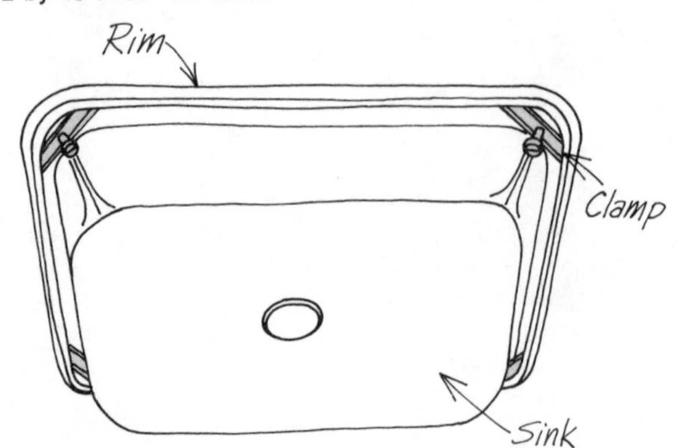

Special clamps *hold rim onto sink. Once sink is installed, other clamps (see page 37) hold it tightly in place.*

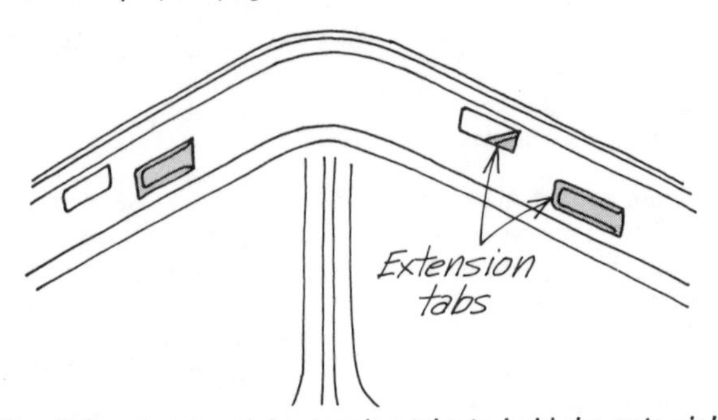

Some rims *have metal extension tabs to hold rim onto sink, eliminating necessity for the corner clamps shown above.*

A self-rimming cast-iron sink rests on a lip on the counter; metal clamps aren't used. Once the faucets, strainer, and trap are disconnected, remove this sink by lifting it out of the counter. To install, coat edge with putty, rest the lip on the countertop and make the plumbing connections.

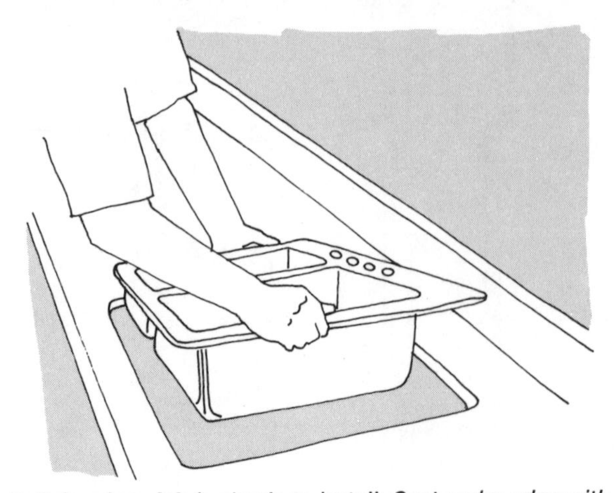

Self-rimming sink *is simple to install. Coat under edge with a ring of plumber's putty, set sink down into place.*

Replacing a kitchen sink strainer

To replace a strainer in a kitchen sink, first remove the J-bend in the trap and then remove the tailpiece that connects with the strainer. The strainer is held tight by a large metal nut (see illustration below). Unscrew this nut with a spud wrench or hammer and screwdriver. Remove the gas kets and lift the strainer out from the top.

Before installing the new strainer, clean thoroughly around the drain hole in the sink. Place a ⅛-inch bead of plumber's putty all around the hole and set in the strainer.

Now you'll need a helper to hold the drain from the top while you fit the gaskets on from underneath and screw on the large metal nut, tightening it by hand as far as it will go and finally using either a spud wrench or a hammer and screwdriver. Wipe off the excess putty in the sink. Connect the tailpiece to the strainer and then replace the J-bend.

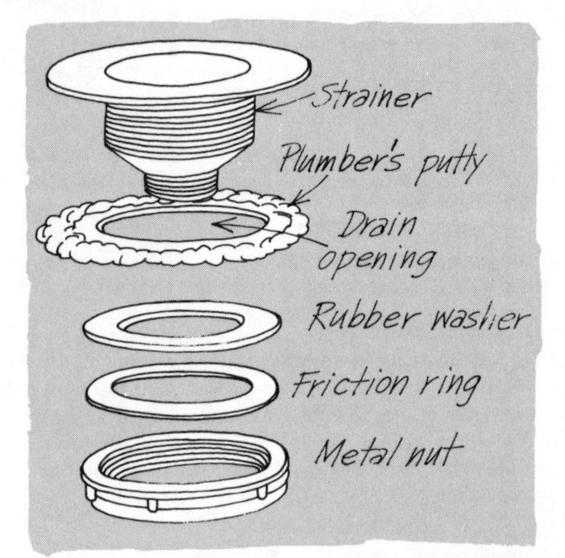

Installing another kind of strainer (shown below) is a one-person operation. All tightening can be done from underneath, with no one needed above to hold it in place.

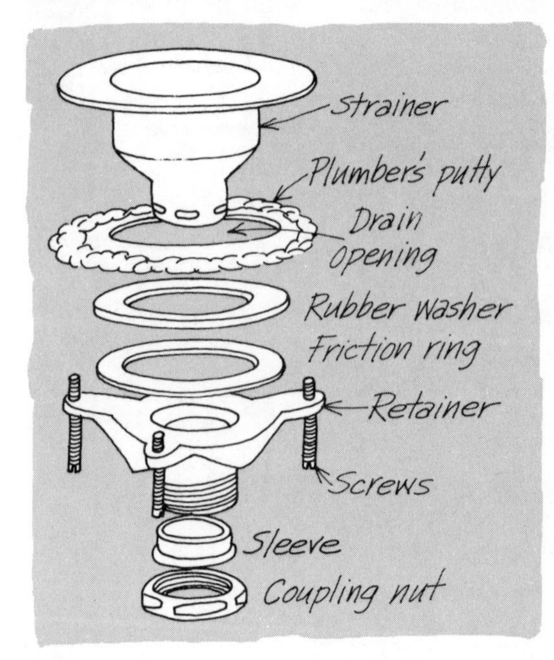

Place a bead of plumber's putty around the hole in the sink and fit the basket into place, pressing it down tight. Put the gaskets on the retainer, fit the retainer onto the strainer, tighten the three screws, and insert the sleeve into the retainer. Wipe off excess putty in the sink. Connect the tailpiece to the strainer and replace the J-bend.

Replacing a bathroom sink

Bathroom sinks may be installed in cabinets just as kitchen sinks are fitted into countertops. They may also hang on a wall, stand on legs, or have a combination of legs and wall support.

To remove a sink. First turn off the water supply to it; then disconnect the drain and disconnect and lift off deck-mounted faucets.

• *Remove a cabinet-top sink* in the same way described previously for removing a kitchen sink.

• *Wall-hung sinks* are mounted on a special hanger. The hanger is screwed onto a support that is nailed between two studs, flush with the walls (see diagram). Once all plumbing connections are removed, the sink will lift off the bracket.

Some sinks have a wall bracket and legs. After you've disconnected everything, lift the sink off the bracket. Legs will unscrew, usually by hand.

To install a sink. A bathroom sink will be easier to install than a kitchen sink because of its smaller size. Choose a rimmed or rimless sink here too. Once the sink is in place, attach faucets, strainer or other drain assembly, and trap. Fit the flexible supply pipes onto the faucets.

• *Cabinet-top sinks.* Use the instructions for installing a kitchen sink (pages 36-38).

• *Wall-hung sinks.* In replacing one wall-hung sink with another, you'll replace the old bracket with a new one. For more strength, avoid reusing the old holes in the brace when screwing the new bracket into place.

If you're installing a wall-hung sink for the first time, you'll have to remove the part of the wall directly behind the sink. The bracket must be attached to a support flush with the unfinished wall, and it must run between two studs. Use a piece of 2 by 6 or 2 by 8. Notch the studs as shown on page 53 and nail or screw the brace into place. Finish off the wall and screw the bracket into place—be certain the bracket is on straight or else your sink will slant to one side. Hang the the sink; attach faucets, drain assembly, and trap; and attach flexible supply pipes to the faucet's threaded extensions.

(Continued on next page)

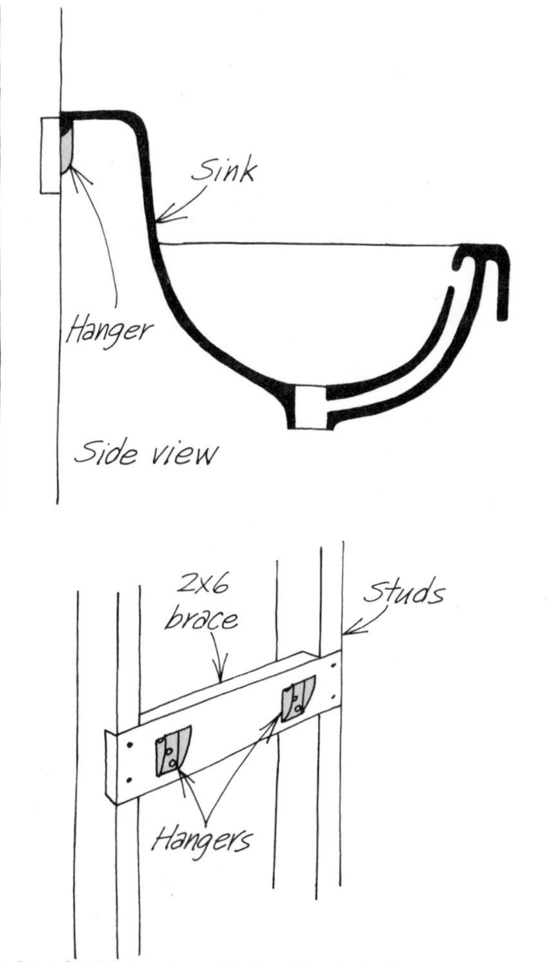

Brace is nailed *into place flush with studs. Hangers screw on. Cut holes for them or screw in through finished wall.*

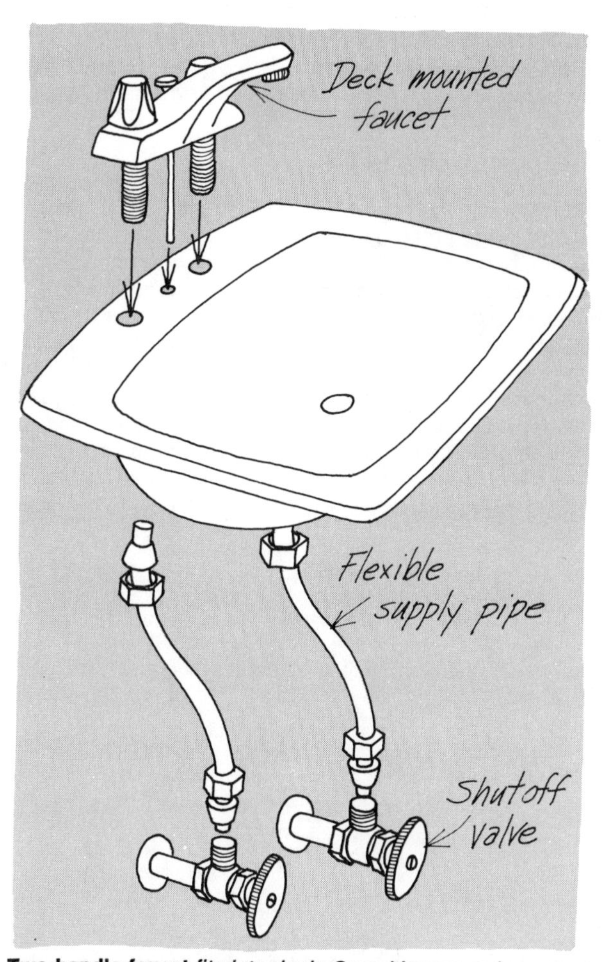

Two-handle faucet *fits into deck. Once it's secured, connect the faucet with flexible supply pipes.*

Some sinks have legs for extra stability; the legs are adjustable to different heights. Use a level and screw the legs in or out until the top of the sink is perfectly horizontal.

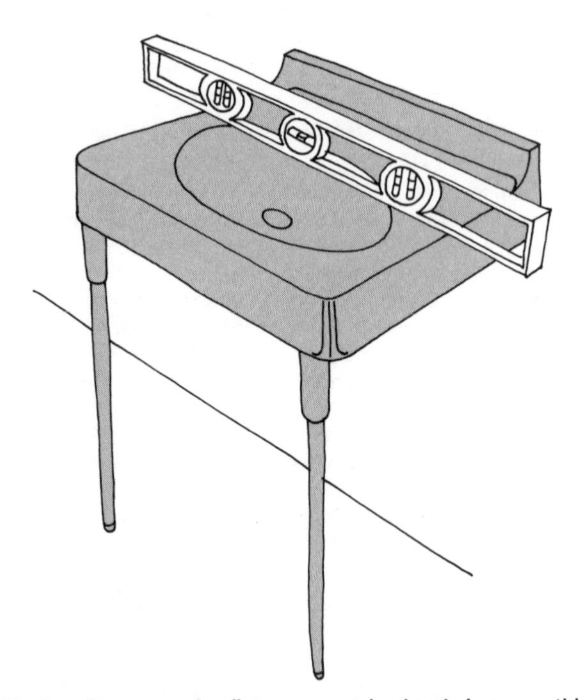

To function properly, *fixtures must be level. Legs on this sink screw and unscrew for making adjustments in height.*

To replace a sink trap

Plumbers say that one of the most common of all household plumbing problems is a leaking P-trap under the sink, caused by corrosion of the chrome-plated brass pipe.

Replacing the trap is a simple task. You can replace just the leaking piece or all the pieces in the assembly. Plumbing supply stores have kits that include the three pipes, washers, and slipnuts. Chrome-plated brass is most commonly used. Plastic is often used when the trap is hidden from view. Specify the size when buying: kitchen sinks use 1½-inch pipe; bathroom sinks use 1¼-inch pipe.

As the diagram below shows, traps have three pieces: a straight pipe, the tailpiece that connects with the sink strainer; then a bend called the J-bend; and finally the trap arm, which connects the J-bend with the drain outlet at the wall. Leaks caused by corrosion almost always occur in the J-bend. To replace this piece, unscrew the slipnuts at A and B and pull the bend off. When buying a replacement, buy new rubber washers as well. Place slipnuts and washers as shown, fit the new pipe into place, and tighten the nuts.

If you're going to replace all three pieces, first loosen the nuts at C and D and remove the assembly. The pipes allow for varying lengths, but if any of the new ones is too long, you can cut it easily with a hacksaw. You can also use an inexpensive pipe cutter for this job—it makes a much neater cut than a hacksaw does.

If you hold the pipes in a vise for cutting, be sure to protect their finish by covering the jaws of the vise with electrician's tape.

Screw the tailpiece onto the sink strainer; then screw the trap arm onto the wall outlet. Put the J-bend on last, fitting it as far onto the other two pipes as it will go. Make certain that the trap arm slants away from the trap.

A double sink may use either of the waste assemblies shown below, called "continuous wastes." These pipes lead to one trap which functions like the trap in a single sink.

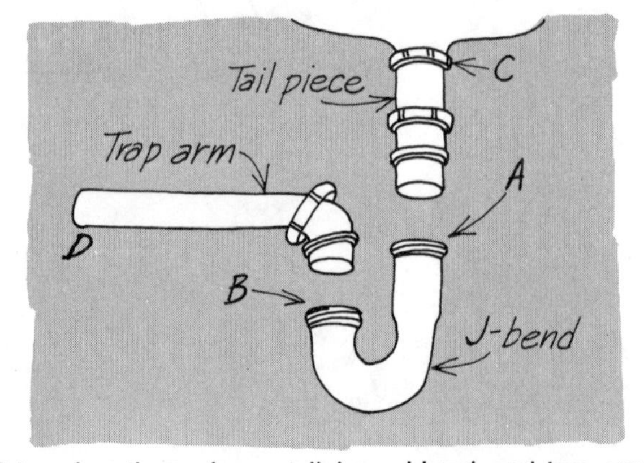

P-trap *has three pieces: tailpiece, J-bend, and trap arm. Replacement tailpiece or trap arm may need shortening.*

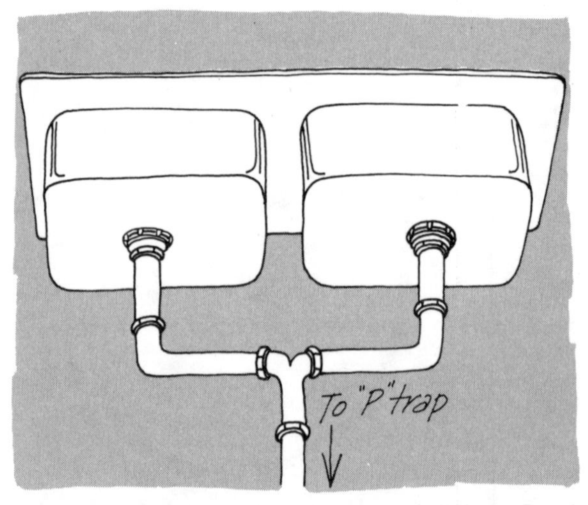

Double kitchen sink *uses "continuous waste" drain. Special fitting joins these drains in center, above trap.*

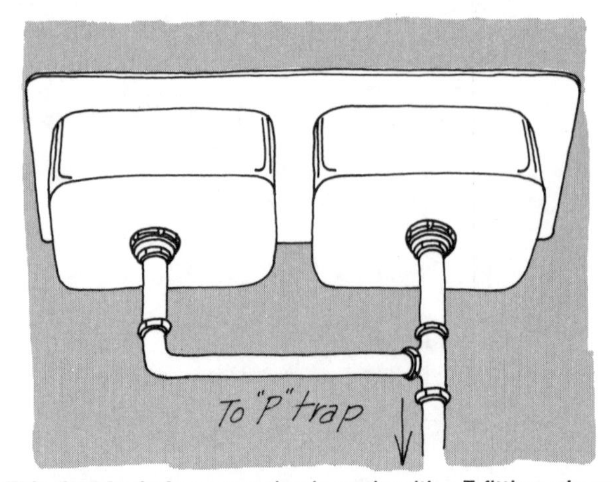

This double drain *connection is made with a T fitting, above the trap. (A disposer would go under the left-hand sink.)*

REPLACING A BATHTUB

Removing a built-in bathtub is a major chore that requires lots of muscle and a good bit of demolition. You'll have to tear out the walls (usually three of them) that enclose the tub. Lifting a cast-iron tub may require three or four people.

When removing a built-in tub, first take off any fixture hardware connected to it—probably just the drain and the overflow cover (see next page), and the spigot. The spigot is just screwed on. To remove it, simply unscrew it, but be prepared for a struggle. Corrosion may have joined the spigot to the outlet very tightly.

If there are tiles along the wall or on the floor that come right up to the tub, chip them out. Now tear the tile or sheetrock from the three enclosing walls. If yours is a steel tub, there may be nails or screws at the very top of the flange helping to hold it in place. Remove them.

With everything disconnected and the walls down, you're ready to lift the tub out. You may have to remove the door to get it out of the room.

Bathtubs are usually made of enameled cast iron or enameled steel. Some manufacturers also produce fiberglass tubs and tub/shower combinations. The extremely heavy cast-iron ones are more durable and generally more expensive. Steel is somewhat lighter, and fiberglass is quite light, making it very easy to install.

Conventional cast-iron and steel tubs have flanges running along three sides. The flanges fit onto horizontal braces in the wall to give added stability once the tub is in place. On some tubs the flanges have holes for screws or nails to secure the tub to braces or studs.

The standard size for new tubs is 5 feet long. If the opening you have is not long enough, you might have to move a wall; you can do this if it's an interior, nonsupporting wall. If the opening is too long, you can build a ledge from the back wall flush with the end of the tub. When buying the tub, keep in mind which is to be the wall side and which the open side.

If the steel tub you install doesn't have holes in the flange for screws or nails, you can drive roofer's nails into the studs just above the flange so the head of the nail will catch the flange and help hold it in place (see illustration).

Cast-iron tubs don't need to be secured with screws or nails. Their great weight makes them very stable, with no need for that extra support. The problem, of course, is lifting the tub to get it into place. Again, take the time to plan very carefully. Three or four people will be needed to lift it.

Secure a fiberglass tub by nailing it in place. Use roofer's nails and nail right through the fiberglass lip onto the floor and onto wall studs.

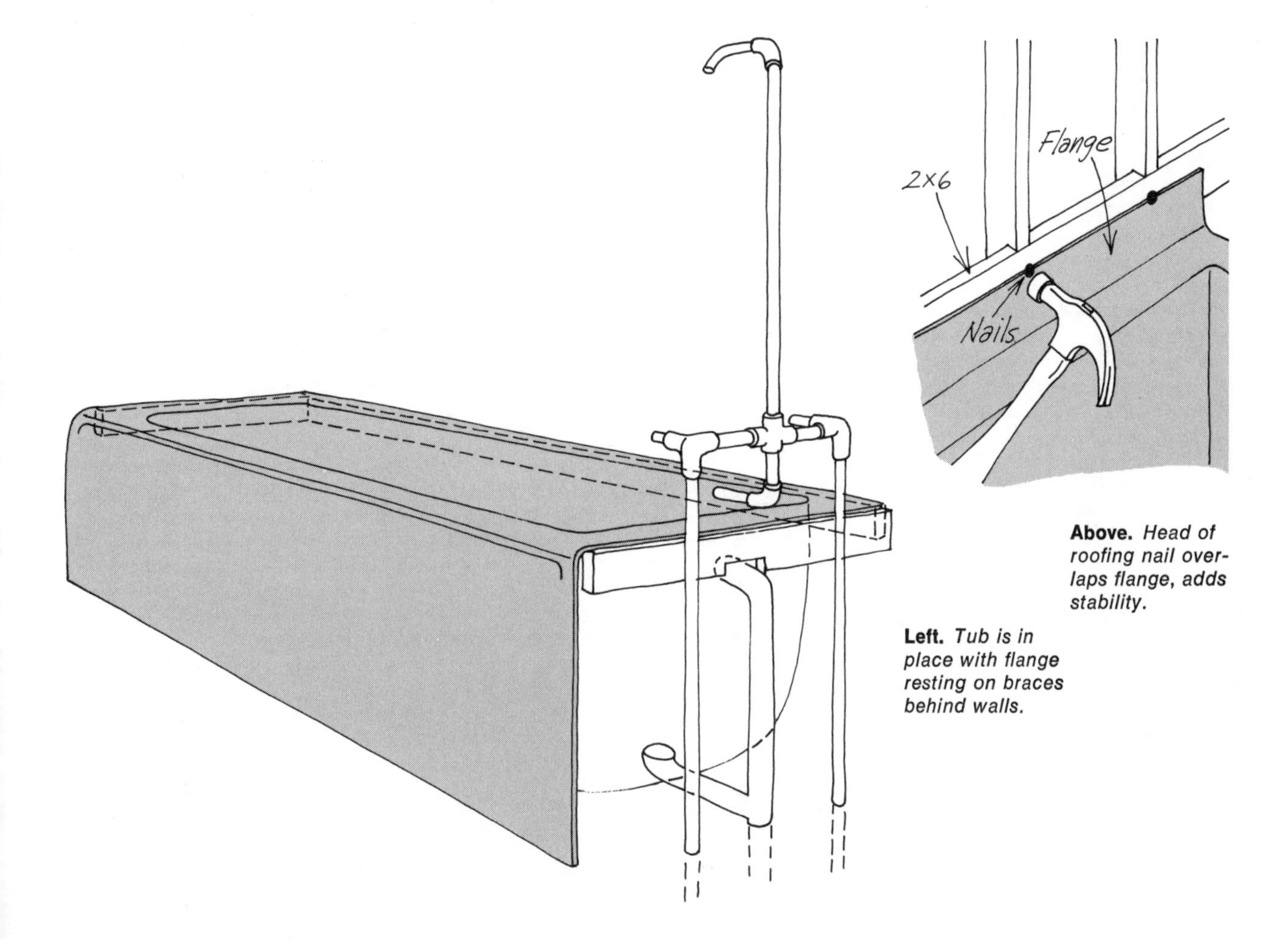

Above. *Head of roofing nail overlaps flange, adds stability.*

Left. *Tub is in place with flange resting on braces behind walls.*

REPLACING A BATHTUB DRAIN

The drawing at right shows how a typical bathtub drain assembly fits onto the tub and connects with the drainpipe. To change this assembly, you have to get at the internal plumbing behind the wall and under the floor.

The strainer in a bathtub drain assembly screws in from the top, from inside the tub. First remove this strainer; be prepared for tough going. There's a special tool designed to do this job, but it's expensive. Try using a hammer and chisel, pounding counterclockwise to loosen the strainer. Then remove the overflow cover. Some of these screw into the drain assembly as a piece. Others have one or two screws. If the assembly includes a lever for opening and closing the drain, disassemble it. It's usually held together by a cotter pin you can get at after you've removed the screws from the overflow cover.

Once the strainer, overflow cover, and lever have been disassembled, the drain itself will come apart easily.

When installing the new drain, be certain all washers are in place before you tighten the slipnuts. Place a bead of plumber's putty around the drain hole inside the tub before you screw the strainer in. Screw the strainer down tight; wipe excess putty from around it.

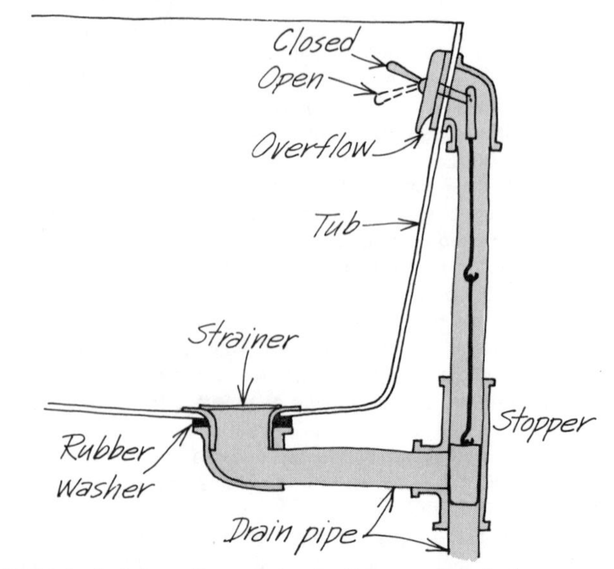

Bathtub *drain/overflow assembly. Excess water drains away through overflow. Lever opens and closes drain.*

REPLACING A TOILET

To remove an old toilet and install a new one, follow the instructions on pages 17-18.

Standard toilets are built so that the center of the bowl outlet is 12 inches from the finished wall. Some are 10 inches, some 14 inches. The odd sizes are usually more expensive. Before buying a new toilet, measure the distance from the center of the drainpipe to the wall.

If the floor flange is rusted, cracked, or broken, you may have to install a new one before you put the new toilet in. The flange may be leaded onto the closet bend. Use a ham-

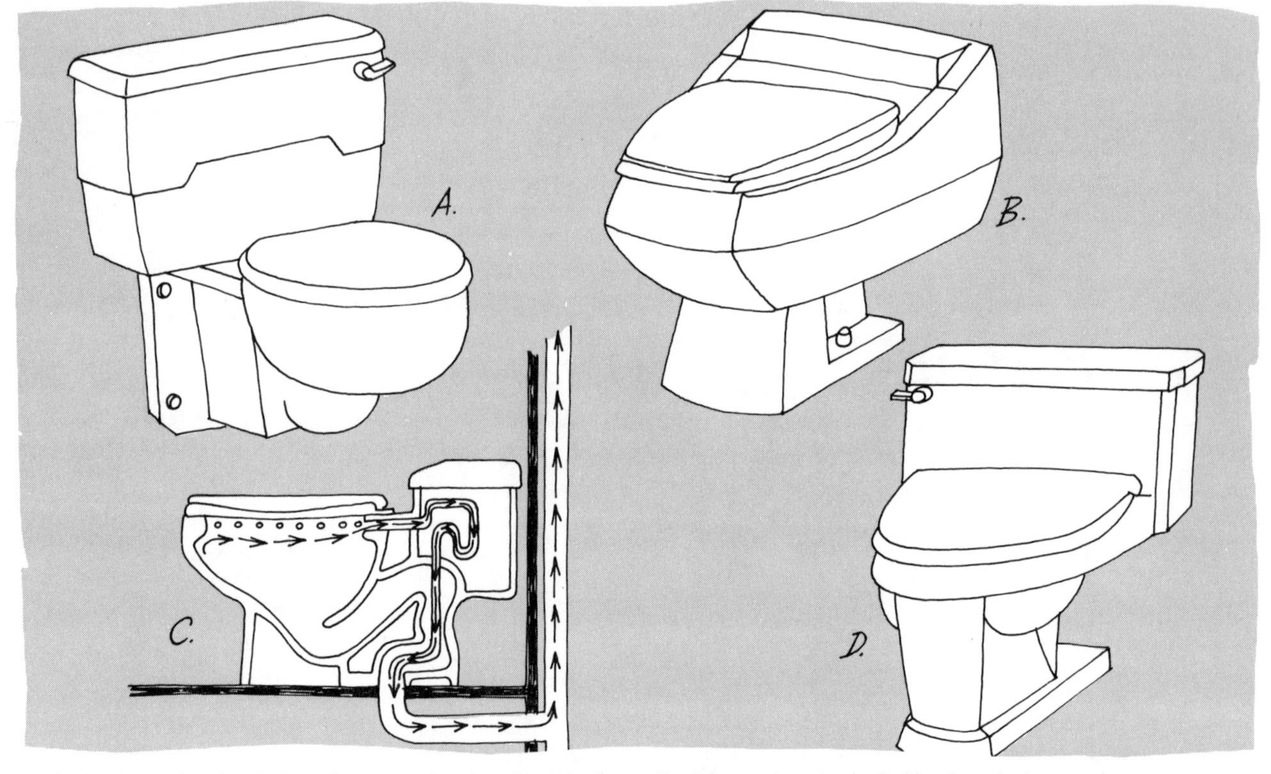

A. *Wall-hung toilet is off floor for easy cleaning.* **B.** *Low-line toilet has modern look.* **C.** *Venting device carries odors through flushing rim into vent pipe.* **D.** *Tank and bowl are one piece, eliminate leaks.*

mer and a cold chisel to loosen the lead. If the closet bend is copper, the flange is soldered and you'll need to unsolder it with a small torch. Remove the screws that hold the flange to the floor and then lift the flange off. When buying a new flange for a lead closet bend get the kind that doesn't need caulking.

Many homeowners choose one-piece toilets when modernizing their bathrooms. If the old toilet has a wall-hung tank, remember when planning that you'll have to remove the bracket if the new toilet doesn't use a wall-hung tank. Unscrew the bracket and fill the holes with wood filler. Sand and finish the wall.

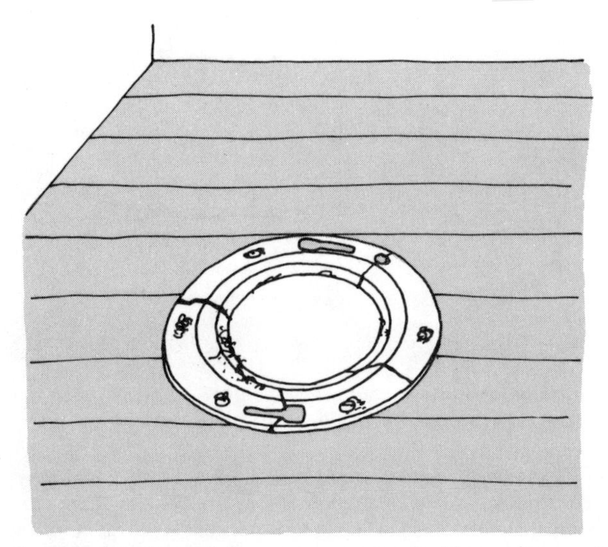

An old, *deteriorated flange can cause leaking at base of bowl. Remove it from closet bend, unscrew it from floor.*

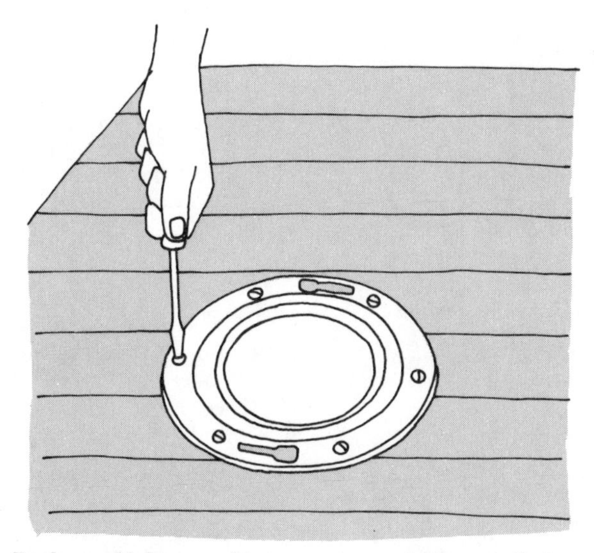

Replace old flange *with a copper or plastic one that can be soldered or cemented into place.*

Caring for your new fixtures

Stainless steel stays bright and shiny if given light and frequent cleaning with a damp cloth, followed by drying with a soft cloth.

For difficult spots and stains that aren't removed by daily cleaning, try ammonia in water or use a solvent like alcohol, baking soda, vinegar, or turpentine applied with a rag. Follow any of these applications with detergent and hot water; then rinse and dry with a soft, clean cloth.

For a high polish, apply a mildly abrasive cleanser and rub only in the direction of the polish lines of the sink to preserve the original finish. A gentle rubdown for a sink should be done with a cloth, soft brush, or stainless steel pad. On a highly polished, lustrous surface, though, a metal pad is likely to mar the finish. Ordinary steel wool and steel wool soap pads should not be used to clean stainless steel—particles may lodge in the surface and rust.

To eliminate fingerprints on a highly polished surface, apply a commercial glass cleaner, baby oil, or automobile wax. After you remove excess cleaner with a soft cloth, a thin protective film will remain. When marks appear, wipe them away with a cloth containing some of the cleaner.

Porcelain enamel fixtures should be cleaned gently with sudsy water, followed by a rinse and dry. For stubborn stains, fill the sink or tub with hot water and add chlorine bleach or oxygen-based bleach, diluted according to label instructions. Let this stand and soak until the stain can be rubbed off. Full-strength or improperly diluted bleach can damage the surface; so can abrasive cleansers. The cleansers contain gritty substances that eventually wear away the enamel surface, making it more difficult to clean as time goes on. Use a commercial cleaner for toilets.

To remove stains *in porcelain enamel sink, pour bleach in hot water and let soak until stain comes off.*

REPLACING A WATER HEATER

If a water heater needs to be replaced, it's usually because corrosion has caused leaks in the tank. Less frequently, the heating mechanism itself fails.

Replacing your old water heater with a new one the same size, using the same fuel, is the easiest thing to do. You might decide, however, that you need a larger unit for your hot water needs. In this case you might have to make some changes in the supply pipe connections.

Extreme caution is required when you work with a combination of water and electricity. The first thing to do when removing the old heater is to disconnect the fuel source. An electric heater should have a nearby circuit breaker to shut off the current. When the current is off, remove the electrical connection from the tank. On a gas burner, turn the valve to an off position—at right angles to the pipe. Then use a wrench to unscrew the connection at the tank's base.

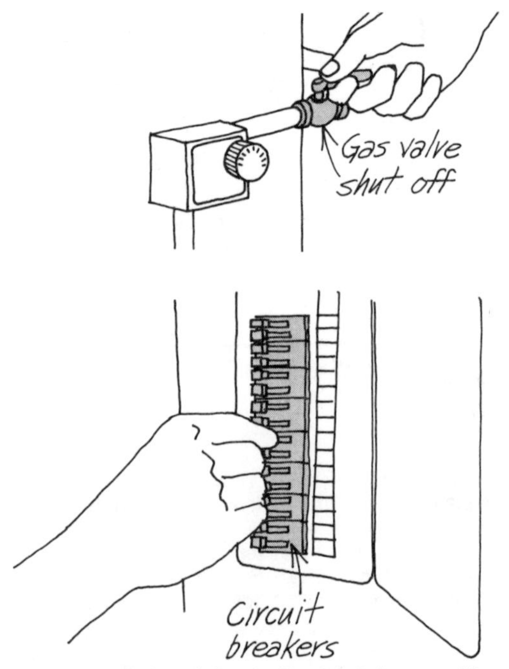

Water and electricity don't mix! *Before working on electric water heater, shut off current at fuse box or circuit breaker.*

Turn off the water going into the heater. If there's no shutoff valve at the heater, turn it off at the gate valve where the water supply enters the house.

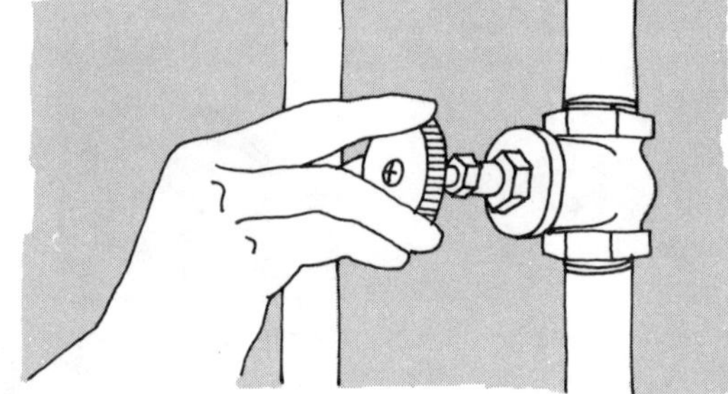

Before opening *drain valve on heater, be certain to turn off the supply of cold water by closing valve on top of heater.*

Drain all the water out of the heater storage tank by turning on the drain valve near the base of the tank. If there's no floor drain beneath this valve, connect a hose to it and run it outdoors or to a nearby drain.

No floor drain *near your water heater? Attach a hose and run it outside onto the ground or to another drain.*

Disconnect the inlet and outlet pipes from the heater. If they have unions on them or if they are flexible connectors with locknuts, the job is simple—just unscrew them. If the heater doesn't have flexible connectors and the pipes are screwed into place without unions on the lines, you'll have to saw through the pipes with a hacksaw (see illustration).

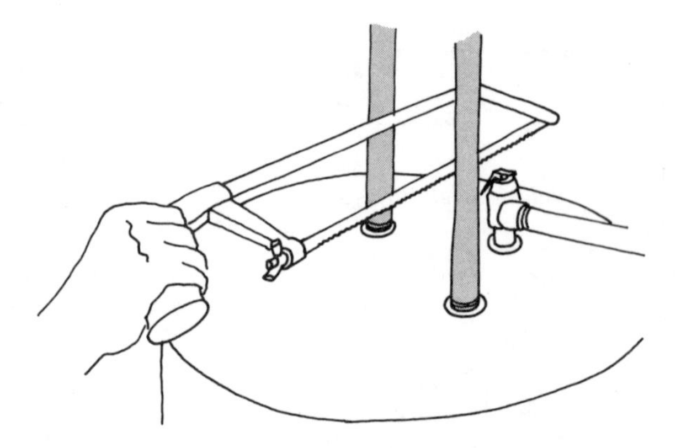

Without special connectors *or a union, galvanized pipes to and from heater must be cut apart with hacksaw.*

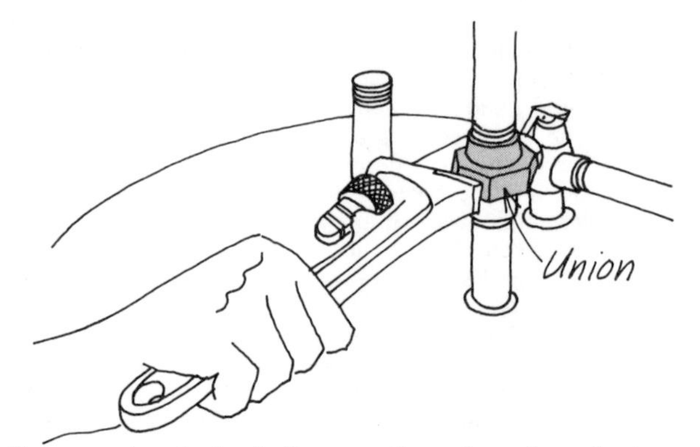

Union *makes task of disconnecting pipes from heater a simple one. Use pipe wrench or large crescent wrench.*

Disconnect the flue from a gas heater by pulling the corrugated male end apart from the smooth female end. Some flues are secured by sheet metal screws as well; these must be removed first.

Many heaters have temperature/pressure release valves to prevent explosions in case the heating mechanism fails. The valves are inexpensive, and it's a good idea to get a new one when you get a new heater.

With the water drained out, all fuel and water pipes disconnected, and the flue disconnected, lift the tank out of place. Once it's out of a confined space, you can set the tank on its side and roll it away.

If the new tank is not the same height as the old one, the old plumbing may have to be tinkered with to make everything come together properly. If you need to add some new pipe, use flexible water connectors, perfect for this purpose. They have threaded locknuts at both ends and simply tighten into place. If your heater doesn't have threaded pipes extending out of the heater at inlet and outlet, you can get two short nipples and screw them into place. Attach the flexible water connectors to these nipples.

Reconnect the fuel supply but don't turn the fuel on. Connect the flue of the gas heater. Next, turn on the water supply and open the hot water tap at a nearby fixture. When water flows steadily from the tap, the tank is full and ready to heat. Now turn the fuel on. Begin by setting the heater at normal and adjust the temperature later if necessary for hotter or less-hot water. Dishwashers require especially hot water; they also increase the demand for hot water.

Flue fits down *over corrugated connection, pulls off easily. Some have screws that must be removed first.*

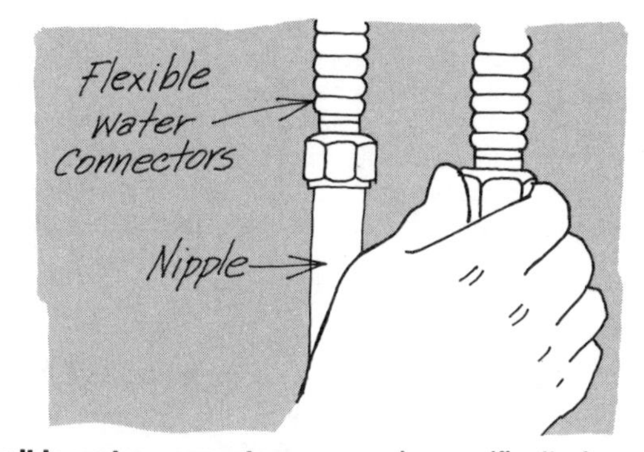

Flexible water connectors *are made specifically for water-heater hookup. Tighten nuts by hand, then with wrench.*

REPLACING A GARBAGE DISPOSER

Always be extremely cautious when working with a combination of plumbing and electricity—watts and water don't mix. Unplug the disposer before working on it.

Working underneath the sink, remove the connection between the disposer and the trap; two screws hold it in place. The disposer comes with its own strainer assembly; a collar fits onto this. The disposer is held in place by being bolted into the collar. Before you unscrew the bolts, put something under the disposer to hold it level while you're unscrewing the bolts and to prevent it from crashing down onto you once the bolts have been removed.

Your new disposer comes with its own strainer, gaskets, and collar. Put a bead of plumber's putty around the drain hole in the sink and fit the strainer into place, pressing down. From underneath, install gaskets and screw on the large bolt with a spud wrench or hammer and a screwdriver. Next, fit the collar on—it may use a metal snap ring to fit into place. Support the new disposer from underneath and attach it to the collar, screwing the bolts on tight. Attach the outlet connection to the trap; it should have a new gasket. Wipe off excess plumber's putty inside the sink. Plug in the disposer.

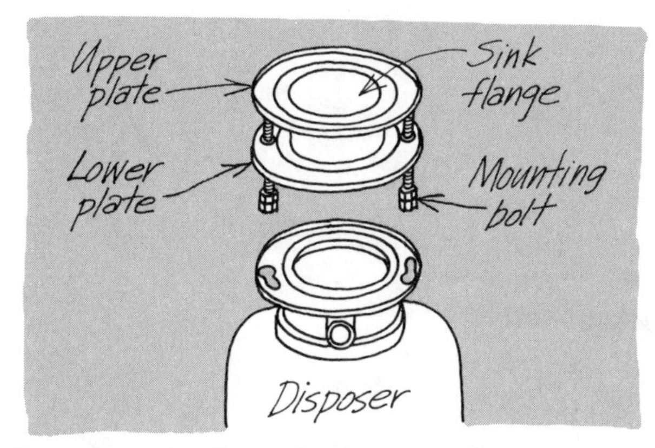

To remove, *take off mounting bolts, turn disposer unit, pull off. Most disposers are mounted in a similar manner.*

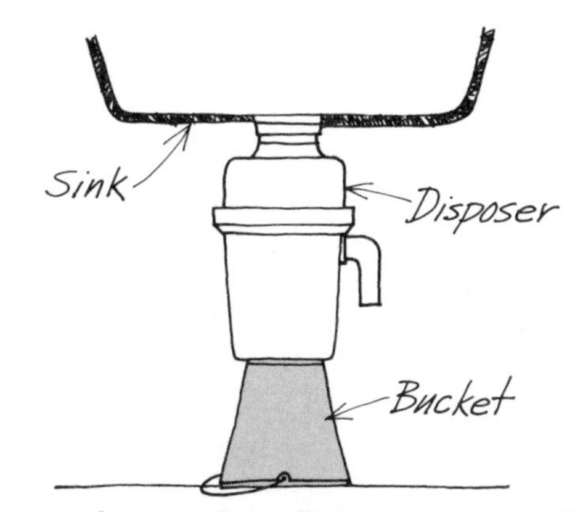

When removing a garbage disposer, *brace it underneath while unscrewing connections; otherwise it will fall.*

Installing new plumbing

Adding new outlets, fixtures, and drains to a house plumbing system can be simple and relatively easy, or it can be quite complicated and difficult. A new washing machine or a new basement shower may require only quick, uncomplicated connections with nearby supply pipes and existing drain pipes. A new bathroom, on the other hand, demands careful and often complex work, both in the planning and the execution of the new plumbing. This section gives you a general guide to the basic steps involved in installing new plumbing. You will need to fit this to your own plumbing system.

The information in this chapter is mostly "what to do." For "how to do," consult the chapter on *Pipefitting know-how,* pages 62-77, and also the chapter on *Replacing old fixtures,* pages 32-45.

START WITH A PLAN

No matter how small the addition you're going to make, a great deal of careful planning should be done before you lift that pipe or tote that wrench. Before you begin, you should know as exactly as you can just what you will do—surprises are more fun at birthdays than they are when you're hefting a section of soil pipe.

Don't buy a pipe, a fitting, or a fixture until you've checked your local plumbing code. (Copies of the code should be available through your building inspector's office. Cost is usually quite low.) Learn what work you may do yourself—some codes require that certain kinds of work be done only by licensed plumbers—and get the necessary permits. Determine what materials and specifications you must use: Is plastic pipe allowed? What are the minimum allowable pipe and fitting sizes? What safety devices must be installed? If you have any questions that the code doesn't answer, ask a building department official.

You'll need to decide how the addition will connect to your existing plumbing system. The task is usually straightforward if the pipes are already exposed, such as in a basement. You'll have to study the entire system, though, to make such a complicated addition as a new bathrom.

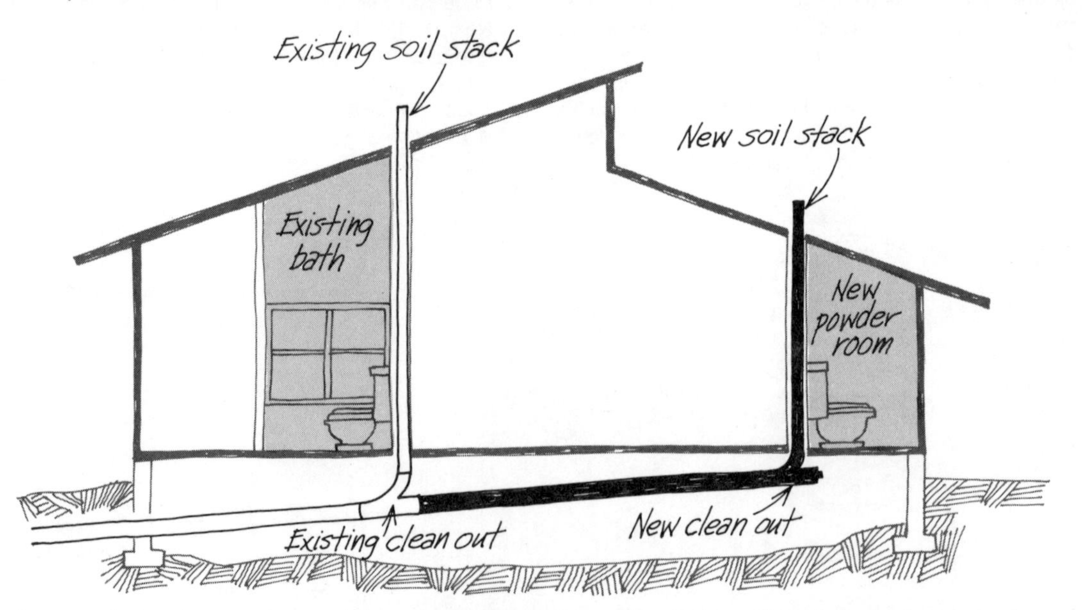

New powder room *is too far away from existing plumbing to hook up with existing soil stack. Instead, new stack, as well as new section of house drain must be installed. See page 49 for a more efficient use of existing plumbing.*

Start in the basement or crawl space with a flashlight, a ruler or tape measure, a pencil and some paper. Make a diagram of the supply and DWV (drain-waste-vent) systems. You need to know what material the pipes are made of (galvanized iron or steel, plastic, or copper), their diameter at the point where you'll be tapping into them, and the distance from the closest pipes to the planned fixtures.

To determine size of pipe:						
Length of string	2¾″	3¼″	3½″	4″	4⅜″	5″
Size of copper pipe	¾″		1″		1¼″	
Size of galvanized pipe		¾″		1″		1¼″

Measure *the pipe's circumference with a piece of string or a tape measure. Then check the table to find pipe size.*

Determine the structural characteristics of the areas you'll be working in. Know where all joists and studs run in relation to the plumbing. (You may have to cut through some of them, but it's good to avoid it if you can.) Measure the clearances you have for working behind walls. (You'll have to cut into an existing pipe run at some point, and available working space is important here.) In your drawing include chimneys, heater ducts, and other obstructions in the walls. Note accessibility to behind-the-wall plumbing, such as any removable panels in closets.

Make a detailed drawing to scale of the location of the planned plumbing additions. Use the drawing to experiment with locations and various possible arrangements of fixtures. For ideas on arranging fixtures in a kitchen or bathroom, see the Sunset books *Planning and Remodeling Kitchens* and *Planning and Remodeling Bathrooms.*

To minimize the cost and keep the work as simple as possible, arrange the fixtures so they're as close to the present pipes as possible. In addition, if there will be more than one fixture, they should share as much of the new piping as possible. The large DWV pipe and fittings are usually much more expensive than supply pipe—something to consider in the planning stage.

Your new fixtures should come with installation instructions and "roughing-in" dimensions. These dimensions indicate where the pipes and the fixture tie together. For instance, most toilets are built so that the center of the bowl outlet is 12 inches away from the finished wall. It's here in the floor, 12 inches away from the finished wall, that the closet bend (the drainpipe that connects the toilet outlet with the soil stack) should be fixed. Make a roughing-in diagram for the new fixtures that pinponts these locations. Some fixtures come with templates you can tack onto floor or wall to mark the roughing-in points.

Decide what kind of pipe you'll use—galvanized iron or steel, copper, or plastic for supply pipes; plastic, copper, or cast iron for DWV pipes. The cost and the ease with which you can work will probably be your most important considerations. (To help you decide, read the chapter on *Pipefitting know-how,* pages 62-77. You don't have to use the same kind of pipe that is already in your system. Adapter fittings will allow you to mix materials.

Finally, prepare a list of the pipes and fittings you need. Your code will specify the minimum diameters required for each use. Normally, supply mains have ¾-inch diameters, branches to toilets and bathroom sinks ⅜-inch. The main drain might be 3 or 4-inch pipe; drains from toilets 2 or 3-inch; drains from sinks, tubs, and showers up to 2-inch—check your code to be certain. Carefully calculate the distance from existing pipes to the new fixtures. Use your drawing when making a shopping list so you don't leave out any fittings or other essential parts (see the next page for a sample shopping list).

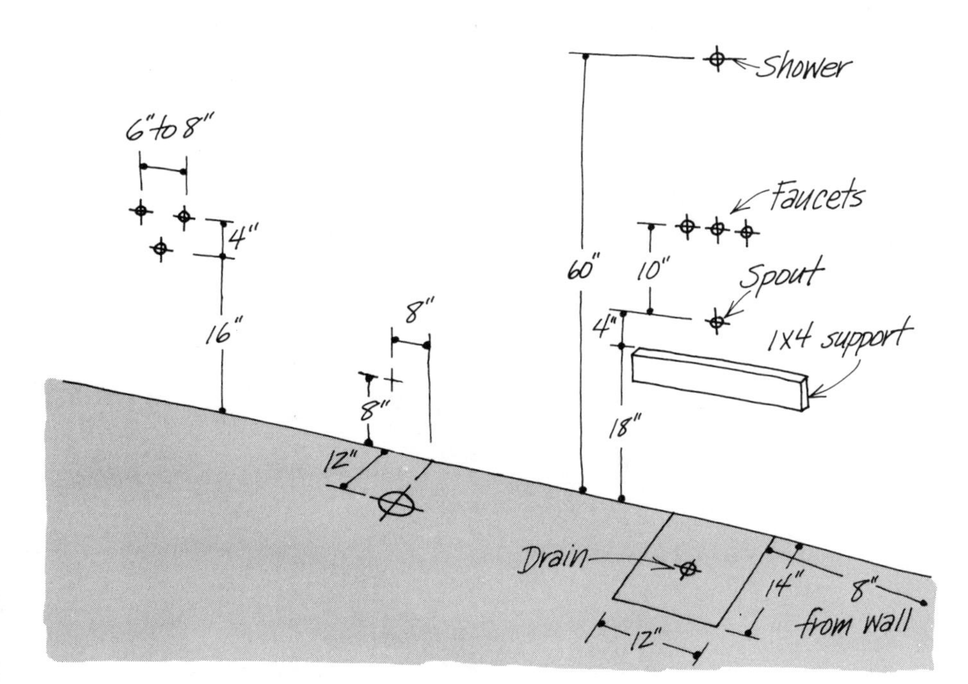

Roughing-in *diagram shows where plumbing connects with fixtures. Measurements will vary with fixtures.*

New plumbing shopping list

Use this sample shopping list for behind-the-wall bathroom plumbing as a check list to follow in making your own list. Check your local code for the pipe and fitting sizes required in your locality.

You'll plan the plumbing for each fixture in three parts: supply, drainage, and venting. Each item is listed in the order it would be installed: the supply system starts from the supply mains (already installed) and runs toward the fixture. The drainage system goes from the drain outlet toward the soil stack. The venting system starts at or near the fixture and goes toward the soil stack. The chart on pages 66-67 will help you plan which fitting to use.

SINK

Supply system

1. 2 reducing Ts (¾" to ½")
2. X ft. ½" pipe
3. Two ½" 90° elbows
4. X ft. ½" pipe
5. Two ½" Ts
6. Two 12" lengths ½" pipe (air chambers)
7. Two ½" caps

Drainage system

8. One 2" x 1¼" x 1½" sanitary T
9. X ft. 2" drainpipe
10. One 2" 90° sanitary elbow
11. X ft. 2" drainpipe

Venting system

12. X ft. 1½" pipe
13. One 1½" 90° elbow
14. X ft. 1½" pipe

TOILET

Supply System

15. 1 reducing T (¾" to ½")
16. X ft. ½" pipe
17. One ½" 90° elbow
18. X ft. ½" pipe
19. One ½" T
20. One 12" length ½" pipe (air chamber)
21. One ½" cap

Drainage system

22. 1 closet bend
23. One 4" sanitary T
24. X ft. 4" soil pipe
25. One 4" combination Y and ⅛ bend (for joining drain under house)

25a. One 4" x 2: sanitary T, tapped double

Venting system

26. X ft. 4" soil pipe
27. One 4"x 1½" cross
28. X ft. 4" soil pipe
29. 1 flashing assembly (for roof vent)

SHOWER

Supply system

30. 2 reducing Ts (¾" to ½")
31. X ft. ½" pipe
32. Two ½" 90° elbows
33. X ft. ½" supply pipe
34. Two ½" Ts for air chambers
35. Two ½" 90° elbows
36. Two ½" close nipples
37. Two 12" lengths ½" pipe
38. Two ½" caps for air chambers
39. X ft. ½" pipe
40. 1 faucet body

Shower supply system (cont'd.)

41. X ft. ½" pipe (above faucet body)
42. One 90° elbow

Drainage system

43. One 2" shower drain/trap assembly
45. One 2" 90° elbow
46. X ft. 2" drainpipe

Venting system

47. 1 reducing T (2" to 1½")
48. X ft. 1½" pipe
49. One 1½" 90° elbow
50. X ft. 1½" pipe

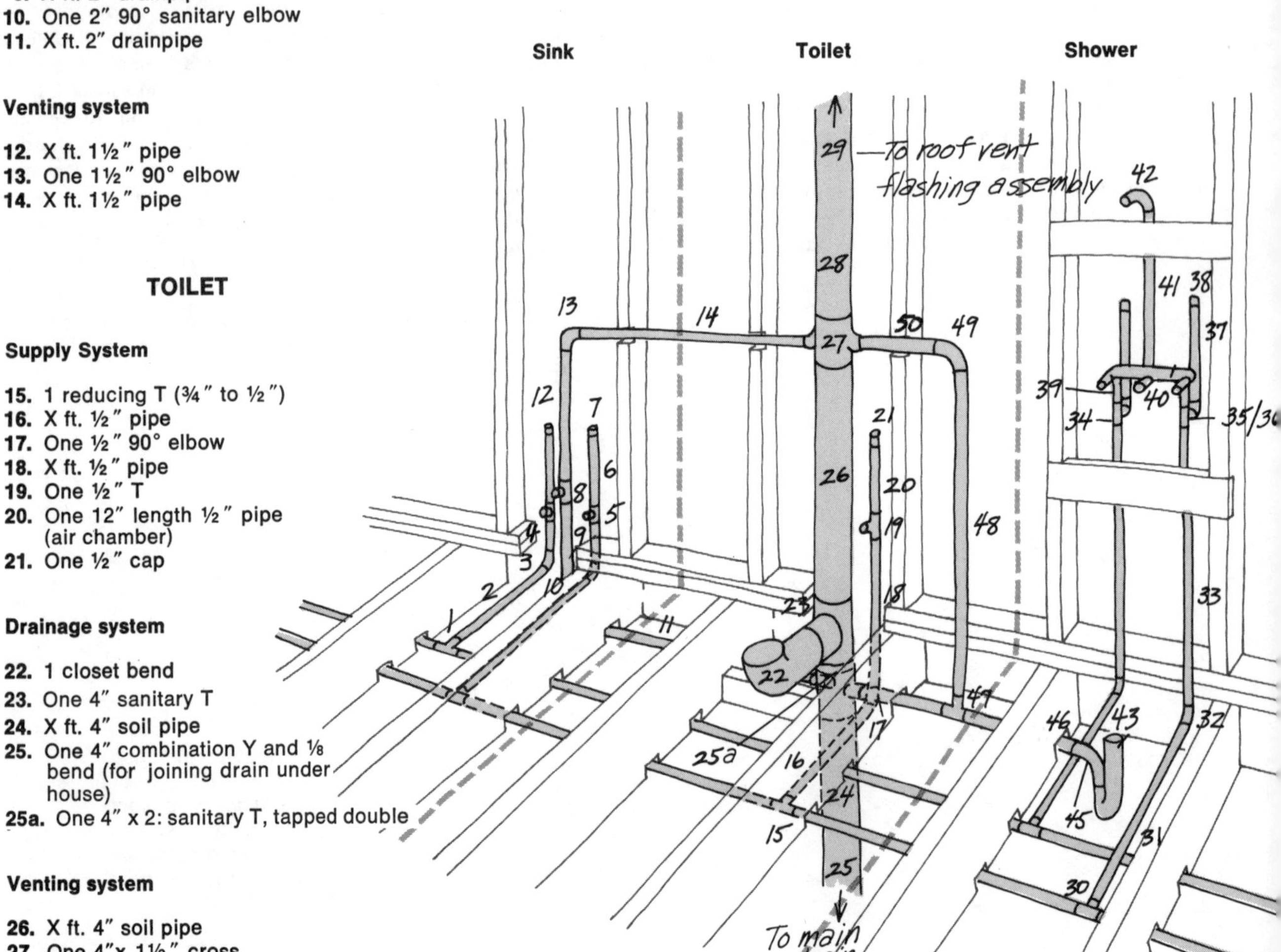

ROUGHING IN A NEW BATHROOM

Here are the steps to follow in the installation of plumbing for a toilet, sink, bathtub, and shower. These basic steps can also be applied to the installation of plumbing in other parts of the house. To do the actual pipefitting, refer to the chapter on *Pipefitting know-how*, pages 62-77.

Roughing-in should always begin with DWV pipes. They're more difficult to handle than supply pipes and aren't as easy to fiddle with after they're in.

Once all the pipes are in place and any required carpentry has been done, you're ready to install the fixtures themselves. Specific information on installing fixtures is in the chapter on *Replacing old fixtures,* pages 32-45.

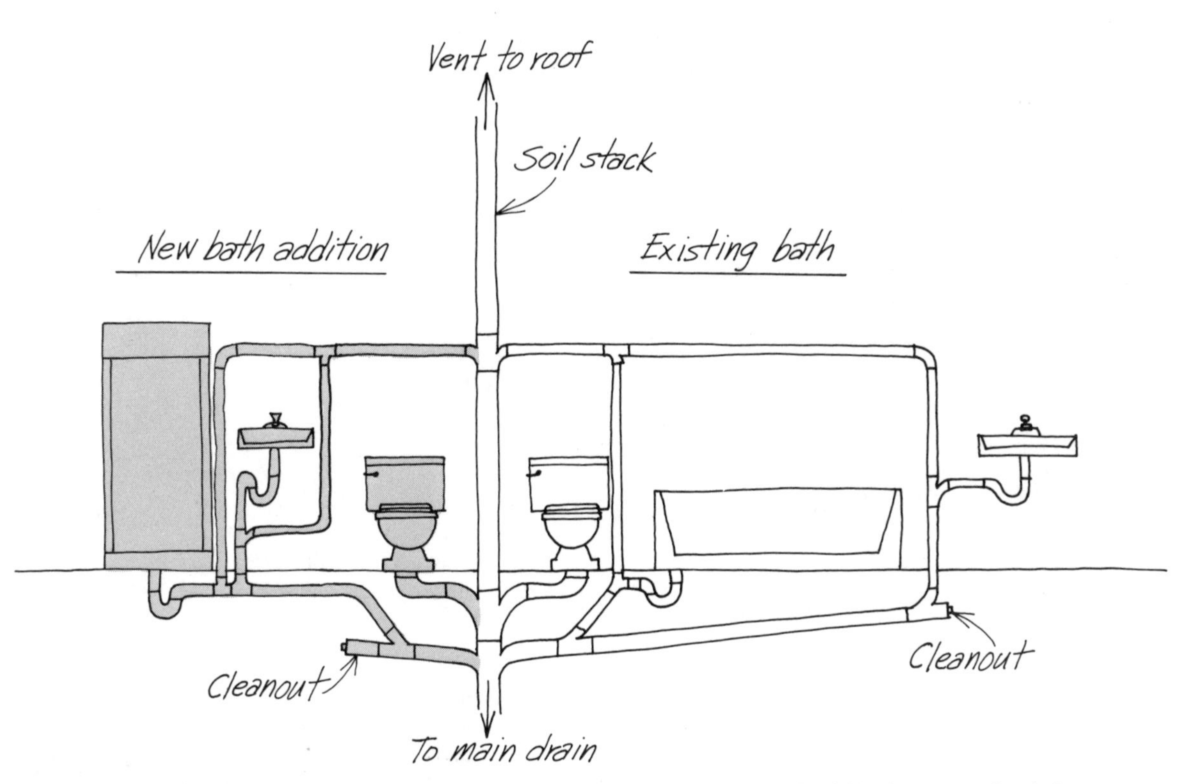

This new bathroom addition, *adjacent to an existing bathroom, is designed to take full advantage of existing plumbing. Savings are substantial when you can tap into the existing soil stack. Also, less drainpipe is needed.*

Two bathroom plans

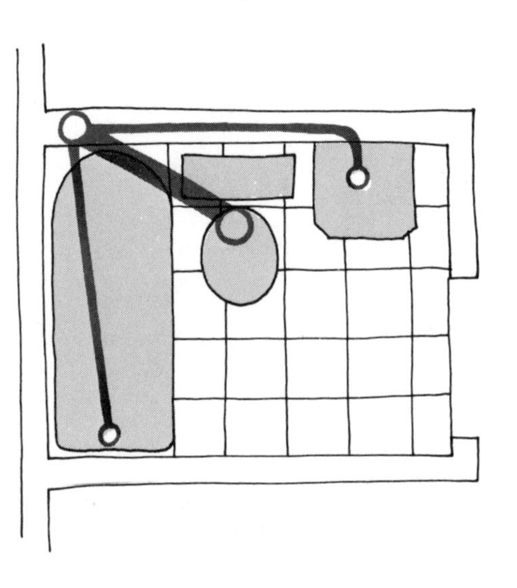

Far left *arrangement uses corner soil stack, minimal amounts of pipe.* **Near left** *is another variation on a three-fixture arrangement.*

Installing DWV pipes

While installing DWV pipes, keep them plugged until the fixtures are connected to them; then the fixture traps will prevent sewer gases from entering the house. (Check with your plumbing supply store about plugs.) Install the soil stack for the toilet first—other drainpipes will connect to it.

Nail a brace into place between joists to support the closet bend and then fit the sanitary tee to the outlet of the closet bend. (If you're tapping into an existing soil stack, add the tee first and then fit the closet bend into it.)

Because lead is a soft metal, it's popular with plumbers for fitting closet bends—it can be bent and hammered into exactly the right positions. Its disadvantages, though, are its weight, which makes lead difficult to work with, and the special equipment and expertise needed to fit a floor panel onto it with oakum and molten lead. Plastic and copper are much easier to work with. The flange is soldered onto the

How to install a soil stack

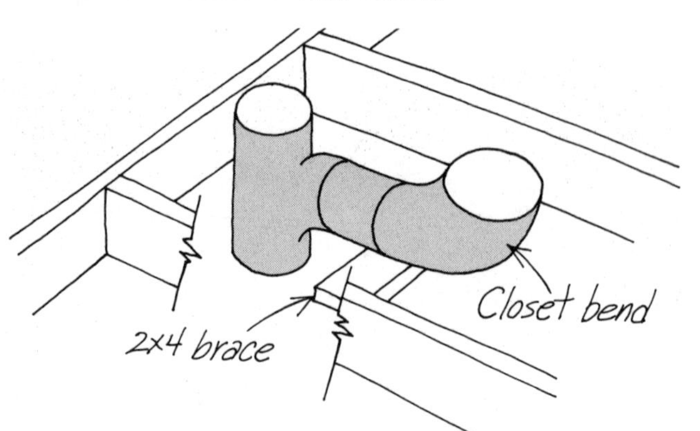

First *fit closet bend and sanitary tee together; then place on a 2 by 4 brace nailed between joists.*

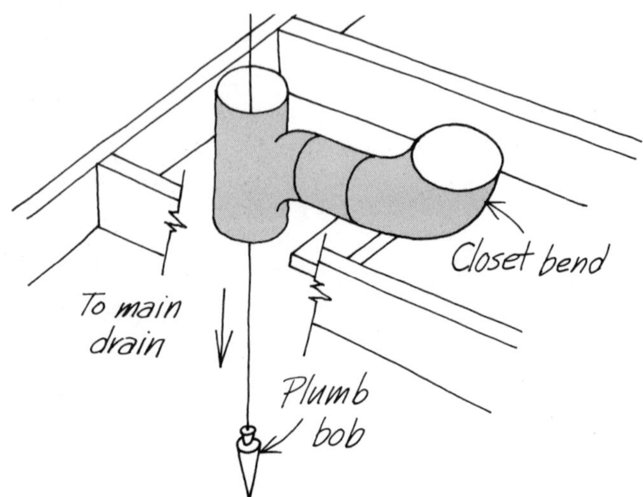

When building a new soil stack, *drop plumb bob through tee down to main drain to make sure stack is straight.*

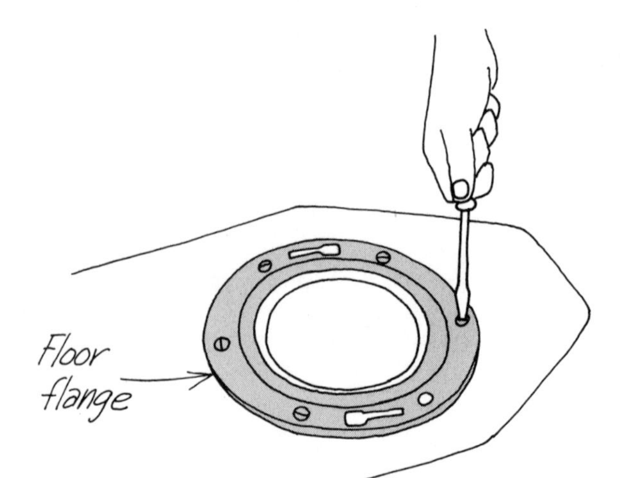

Fit floor flange *onto closet bend first; then screw it onto the floor. Be sure bolts line up with bolt holes on toilet.*

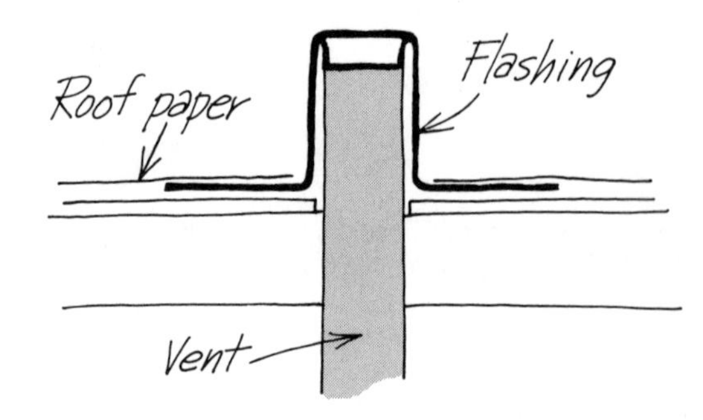

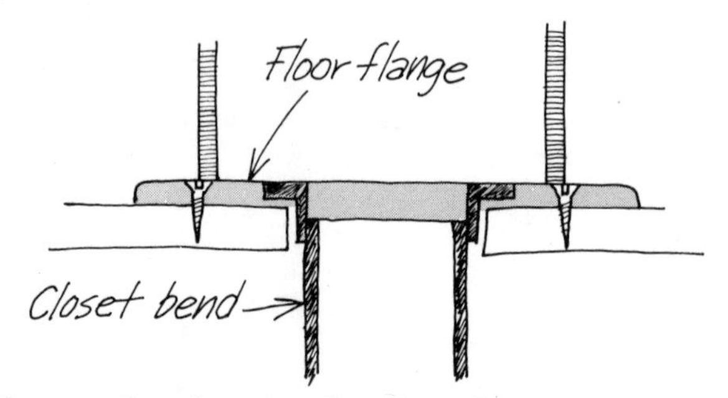

Cross section *shows how floor flange fits onto closet bend. Copper and plastic are easiest to work with here.*

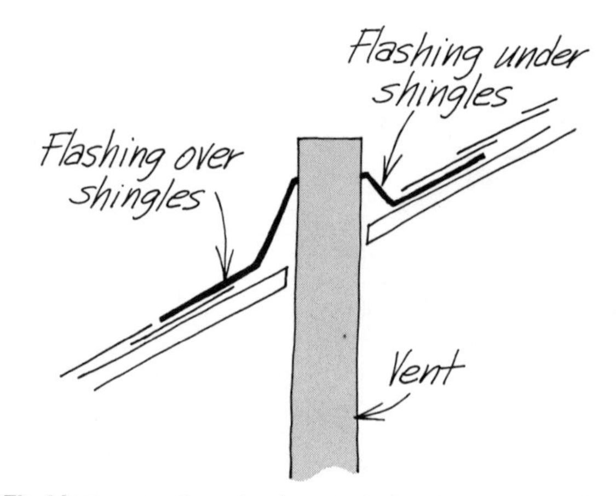

Flashing *prevents water from entering house through hole for vent. On slanted roof, cover flashing on upper side.*

copper bend (solvent-cemented into the plastic one) and then screwed down onto the finished floor.

While someone holds the closet bend exactly in place, drop a plumb line down through the tee (as illustrated) to determine where the soil stack will connect with the drain. Build the stack up from the drain to the tee and then on up to the roof. (This sequence could be reversed in a new building.) Add an increaser if your code requires it.

Flashing must be installed to prevent water from entering the house through the vent opening. Fit the flashing down over the vent pipe. It should fit under shingles or shakes except on the downslope side of a slanted roof. Coat the bottom of the flashing with roofing asphalt, fit it in place, and nail it down. If the nails aren't covered by roofing, dab some asphalt on them to prevent leaking. Also dab asphalt between the vent pipe and flashing to seal the top. Replace the roof covering. On existing roofing, install the flashing first, pushing the vent up through the installed flashing.

Once the closet bend and soil stack are in place, you can install other DWV pipes. Work from the stack back toward the fixtures. Check your building code to determine what types of supporters are required (metal straps are the most common). The drainpipe for the bathroom sink should extend out about an inch beyond the finished wall. The bathtub drain has a trap underneath.

After the drainpipes are in, install the vent pipes. Here again, work from the stack back toward the fixture.

Installing supply pipes

Before you install the supply pipes, turn off the main shutoff valve and drain the supply system by opening the highest and lowest faucets in the house.

Tap into the existing system with tees at the points you've marked on your scale drawing. Build the lines toward the fixture. (Horizontal runs of supply pipe should slope away from the fixtures ¼ inch per foot so that the system can be drained.)

Remember that the hot-water supply pipe will be on the left side of the fixture as you face it, the cold-water on the right. For the sink and toilet, fit a tee onto the run at the appropriate distance above the floor (check the fixture specifications). Extend an air chamber above the sink and toilet (see page 27). Fit a nipple into the tee so that it will extend about an inch out past the finished wall. Shower and tub plumbing takes a faucet body rather than tees. To add air chambers, tap into the supply pipes below the faucets.

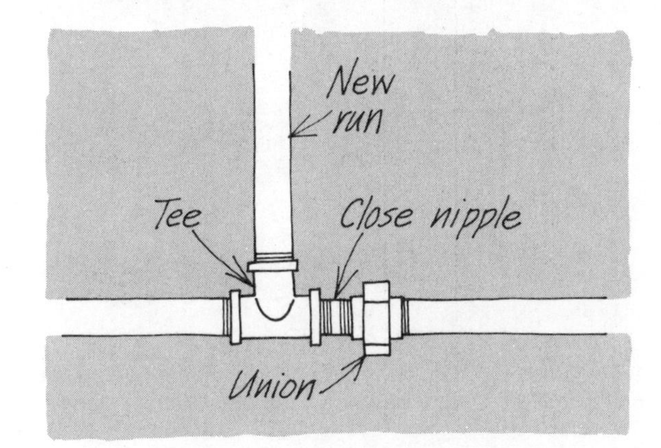

Use a T to tap into an existing run. Rigid piping materials will need a union with the T; flexible piping won't.

CUTTING JOISTS AND STUDS

Try to route new pipes between existing wall studs and floor joists as much as possible. You may find, though, that you'll need to cut through some studs and joists. Follow the suggestions to avoid weakening the structure.

Cutting into floor joists

If a pipe run hits a joist in the center, you can drill a hole through the center of the joist, but the diameter of the hole must be no larger than one quarter the depth of the joist. A hole through a 2 by 8-inch joist should be no larger than 2 inches in diameter. The hole *must be centered* between the top and bottom of the joist.

If the pipe run hits a joist at the top or bottom, a notch in the joist will accommodate it. Notch the joist close to the points of support—the closer the notch is, the less weakening will occur. Make the notch, top or bottom, no more than a quarter of the joist's depth. If the notch is to be within a few inches of the point of support and on the top of the joist, you can make the notch size up to ⅓ the depth of the joist.

(Continued on next page)

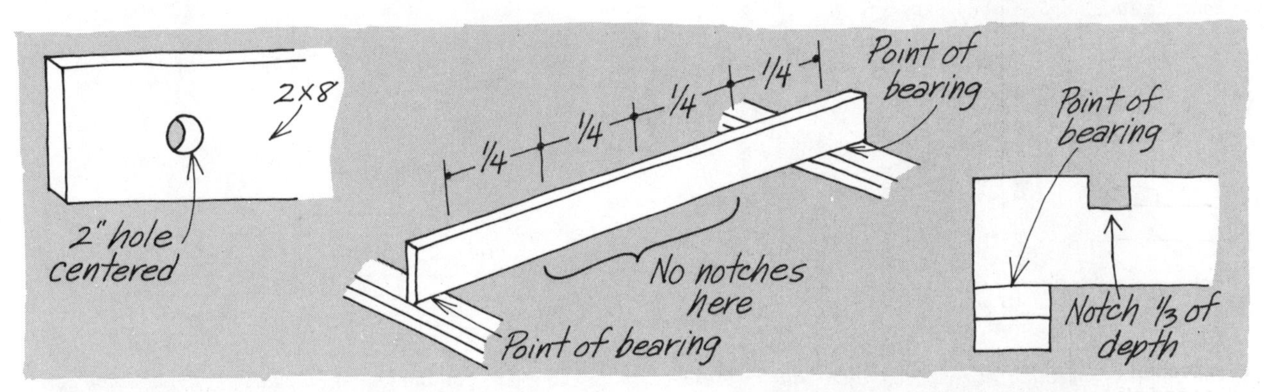

Cutting into joists, *left to right: hole centered on joist; notches only at joist's end; top notch 1/3 of depth.*

Joists notched at the top should have lengths of 2 by 2-inch boards nailed in place under the notch on both sides of the joist to give added support at that point, Joists notched at the bottom should have either a 2 by 2-inch piece of board or a steel brace nailed in place across the notch for extra support.

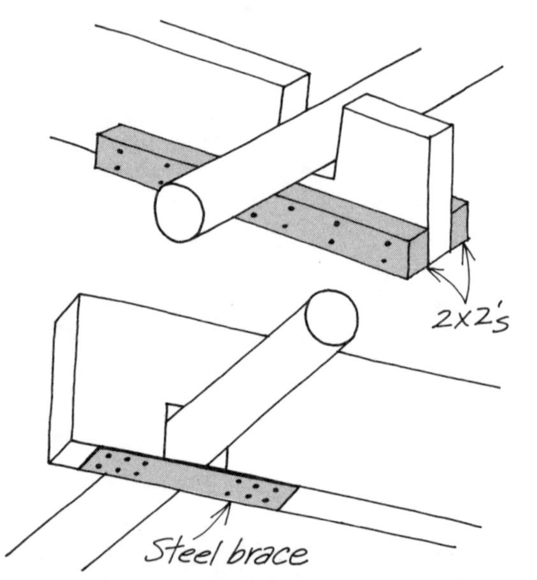

Brace a notch *cut in top of joist below notch on both sides. For bottom-notched joist, use a steel brace across opening.*

You might sometimes need to cut a section out of a joist to accommodate a closet bend, as shown below. Make this cut only in the end quarters of the joist. Reinforce that section by using "headers" on both sides of the cut. Headers are lengths of board the same width and depth as the joist, nailed into place as shown. It's a good idea to use double headers for support if you've removed an entire section of joist.

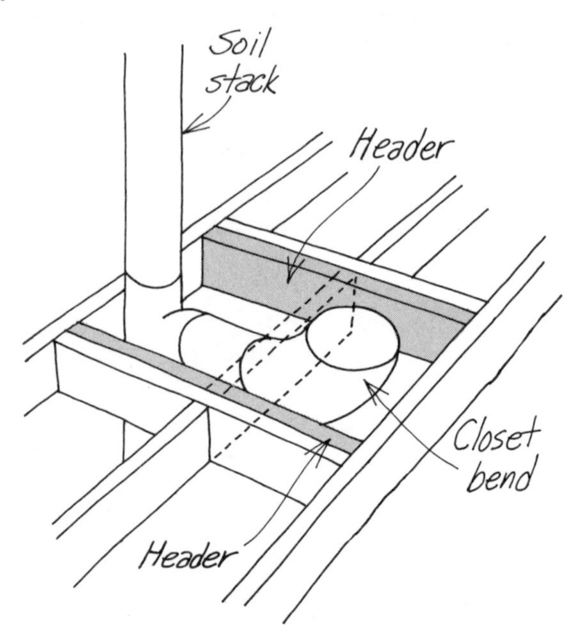

If you remove *a section of joist, reinforce with headers—sections of board the same width and depth as the joist.*

Cutting into wall studs

The diameter of a hole drilled in a wall stud should never be larger than ½ the depth of the stud. Drill the hole in the center of the stud's width.

You can safely notch a wall stud in its upper half up to ⅔ of its depth without reinforcing it. Don't notch the bottom half of a stud more than ⅓ of its depth without reinforcing it. (You may want to reinforce it anyway if you have room for a brace. Use a steel or wood brace—called an FHA plate—nailed in place across the notch.)

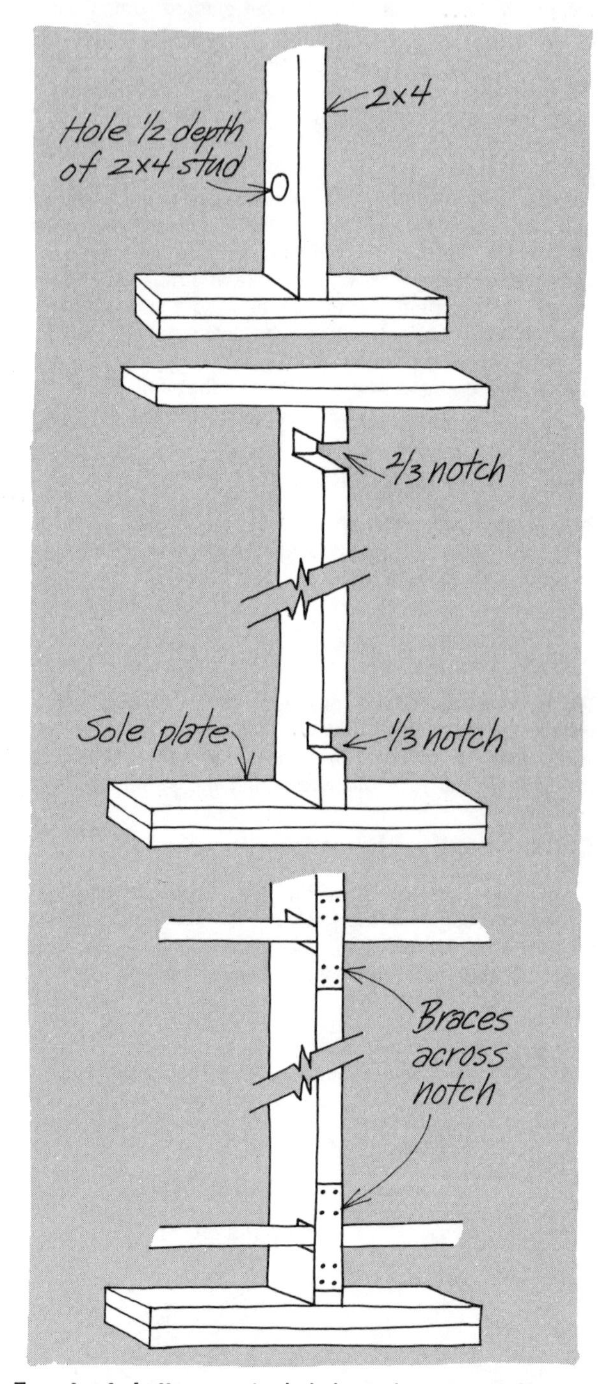

From top to bottom: *center hole in stud; upper notch can be 2/3 stud's depth, lower notch 1/3; nail brace across notch.*

How to cut a notch

Mark stud *with a pencil where notch will go; then use handsaw to cut both sides to the required depth.*

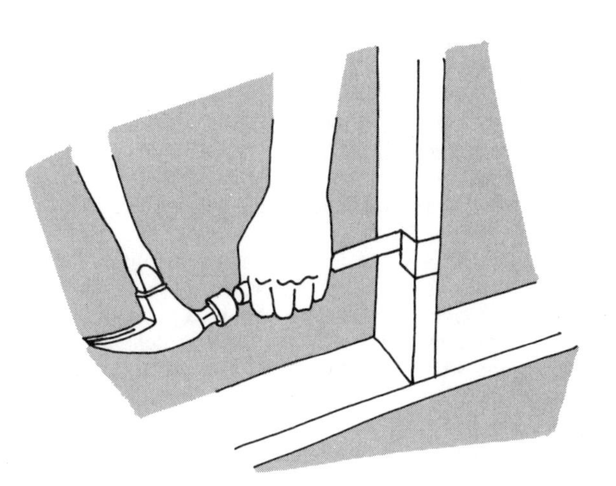

After the edges *of the notch have been cut, use a hammer and sharp chisel to cut away the wood in between.*

PLUMBING FOR A WASHING MACHINE

Adding plumbing for a washing machine is a fairly straightforward task. It's simplified greatly if supply pipes are nearby, such as for a laundry tub. In that case, you can tap into these pipes and run the drain hose into the existing drain.

If you're starting from scratch, you'll need not only supply connections but also a special drainpipe called a "standpipe." Check your local code for the required height of this pipe. As the drawings indicate, some standpipes come complete with trap. These are the simplest to install. Tap into the nearest drain and fit the standpipe to it.

Supply pipes for an automatic washer are usually ½ inch in diameter. Check your local code and also the manufacturer's instructions before you install them. Both the hot and cold lines need air chambers and shutoff valves.

Some manufacturers recommend that air chambers be of pipe one size larger than the supply pipes themselves and that the chambers be as long as 24 inches—1½ to 2 times longer than ordinary air chambers. This extra length provides more cushion for the abrupt automatic turnoff.

The washing machine connects to the water supply with short lengths of hose that have female hose fittings at each end. It's a good idea to install shutoff valves to turn off the water supply when the washer is not in use (use hose bibs for the valves). This shutting off will prolong the life of the automatic valves by eliminating the constant pressure.

The special valve shown below shuts off both hot and cold water simultaneously with just a flick of the switch. It can be installed in place of the existing valves with little or no modification.

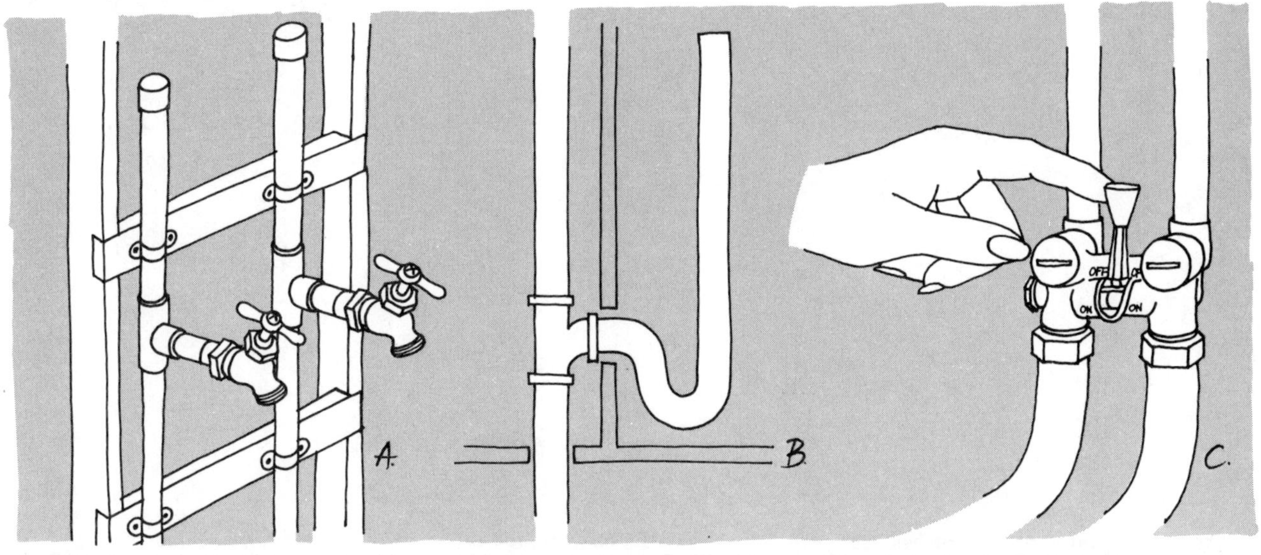

A. *Extra-large air chambers on washing machine supply pipes.* **B.** *Washing machine drainpipe (called "standpipe") can be all one piece.* **C.** *This single valve controls flow of hot and cold water to washing machine.*

INSTALLING A DISHWASHER

When you purchase a new dishwasher, be certain to get the roughing-in dimensions from your supplier. You'll need these to hook the dishwasher up to the hot-water supply.

The examples shown below present the basics of dishwasher installation. Plan to put the dishwasher as close as possible to the sink—that really means as close as possible to the hot-water supply pipe and also to the sink's drain.

Using a tee, tap into the hot-water line and build a run to the intake valve on the dishwasher. Check the manufacturer's specifications to see what type of fitting is necessary at this point. If it's possible, put a shutoff valve on the line so you won't have to turn off the entire supply when working on the machine.

The dishwasher will come with a drain hose that clamps onto an outlet pipe in back. This hose must be free of kinks so the flow of drain water is not restricted. The bend from the outlet down to the floor should be gradual; the hose should be left long enough to allow for a gentle curve.

For your sink drain, you'll need to buy a special drain tee onto which the hose will clamp. Fit it onto the drain above the trap. There's also a special fitting available that allows the dishwasher drain to connect with the garbage disposer. (Ask your appliance dealer about this fitting.) Most codes require that dishwasher drains be fitted with a special air gap or vacuum breaker. This device functions to prevent a vacuum in the waste line that could cause siphoning of waste back into the washer.

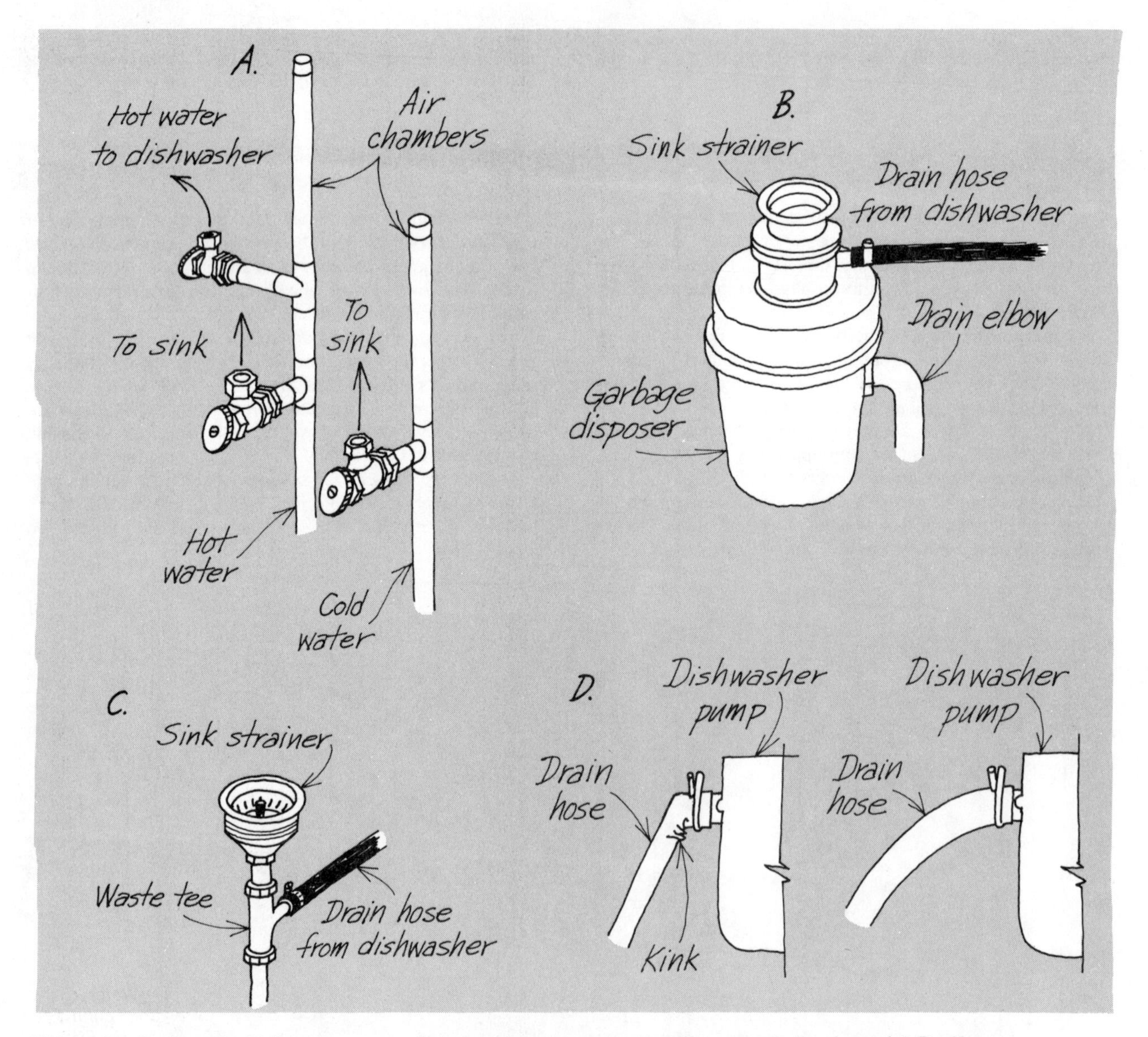

Dishwasher hookup: A. *Dishwasher uses only hot water; a separate shutoff on supply line is useful.* **B.** *Almost all garbage disposers come equipped with fitting for dishwasher drain hose. Hose clamps in place.* **C.** *Special drain tee for dishwasher fits above the trap.* **D.** *Avoid crimping the dishwasher drain hose; make curve gradual.*

An easy-to-use tee

If you want to make a hookup to copper or galvanized pipe without the fuss of cutting pipe and threading or soldering, use a saddle-type tee. All you need to do is bolt the tee into place, drill into the pipe with a hand drill or a power drill, and screw the threaded connection (hose cock, pipe, nipple) into place.

Some saddle-type tees eliminate the need for a drill. They're built to pierce the pipe as they are tightened into it.

This fitting is ideal for adding basement showers, washing machines, or dishwashers. For longer pipe runs, though, it's still best to use standard fittings.

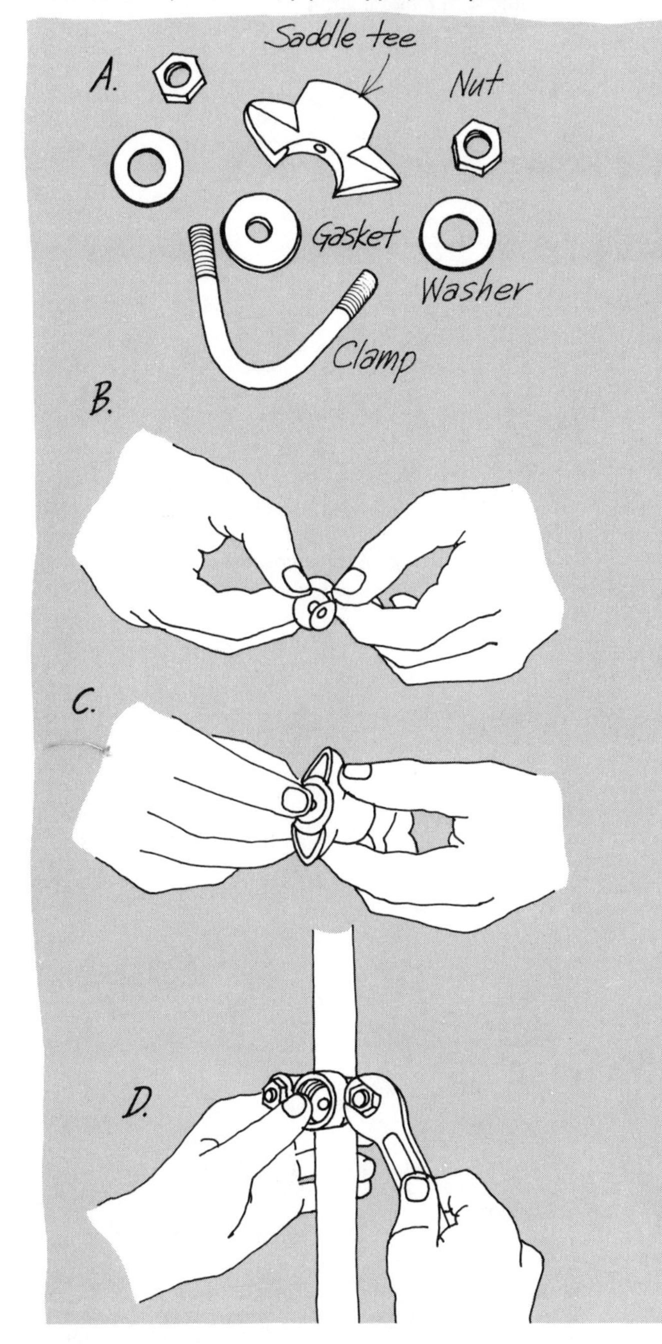

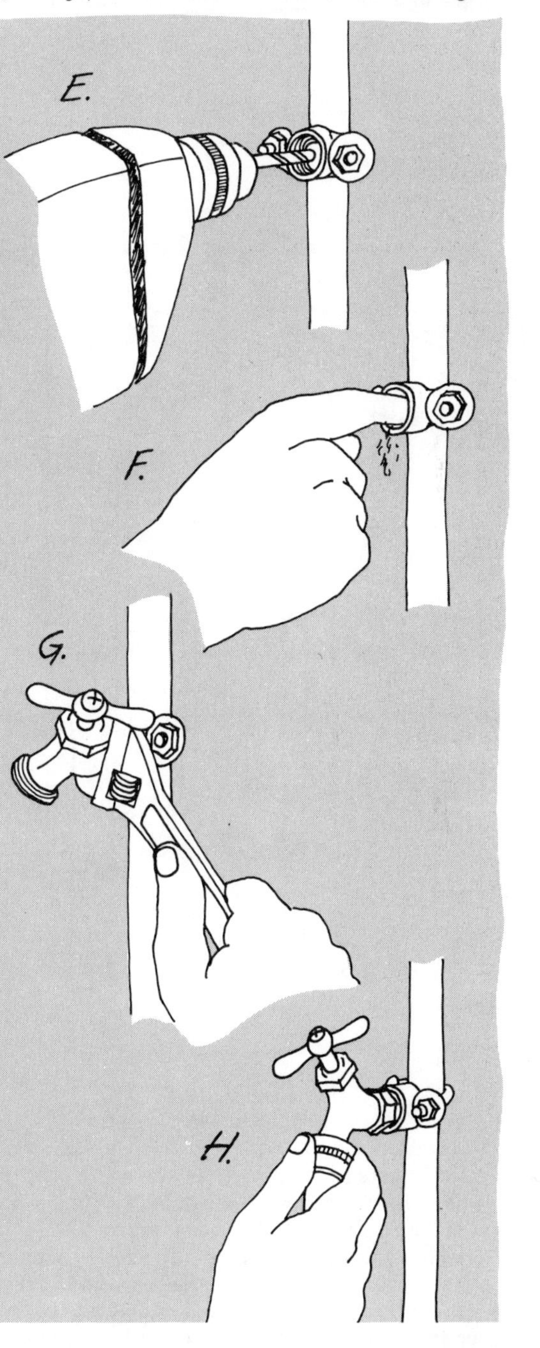

A. *Saddle-tee allows quick tap into supply pipes.* **B.** *Peel sticky backing off of gasket (if it has one.)* **C.** *Fit rubber gasket into place where tee will touch pipe.* **D.** *Fit tee onto pipe; tighten bolts with crescent wrench.* **E.** *Drill through hole in tee directly into pipe. (Water off!)* **F.** *Clean out all little chips and bits of metal.* **G.** *Fit connection into tee (tees come threaded or smooth).* **H.** *Finish hookup, turn on water supply, and check for leaks.*

ADDING A BASEMENT SHOWER

It's very little trouble to add a shower to your basement if there's already a floor drain in place. The addition of the drain though, is difficult and time-consuming. A simple situation in which hot and cold supply pipes run horizontally overhead is illustrated below. Directly beneath, in the cement floor, is a drain.

Tap into both lines with a tee. In the case shown, the hot-water line comes straight down, right next to the wall. The cold-water line uses the tee, then a nipple, then a 90° elbow, another nipple, and another 90° elbow to get to the wall. Pipes can be braced to the wall with steel straps.

Fit the faucet body and shower head onto the pipes. Then fix a short length of 2 by 4 to the wall as shown and secure the pipe with a metal strap.

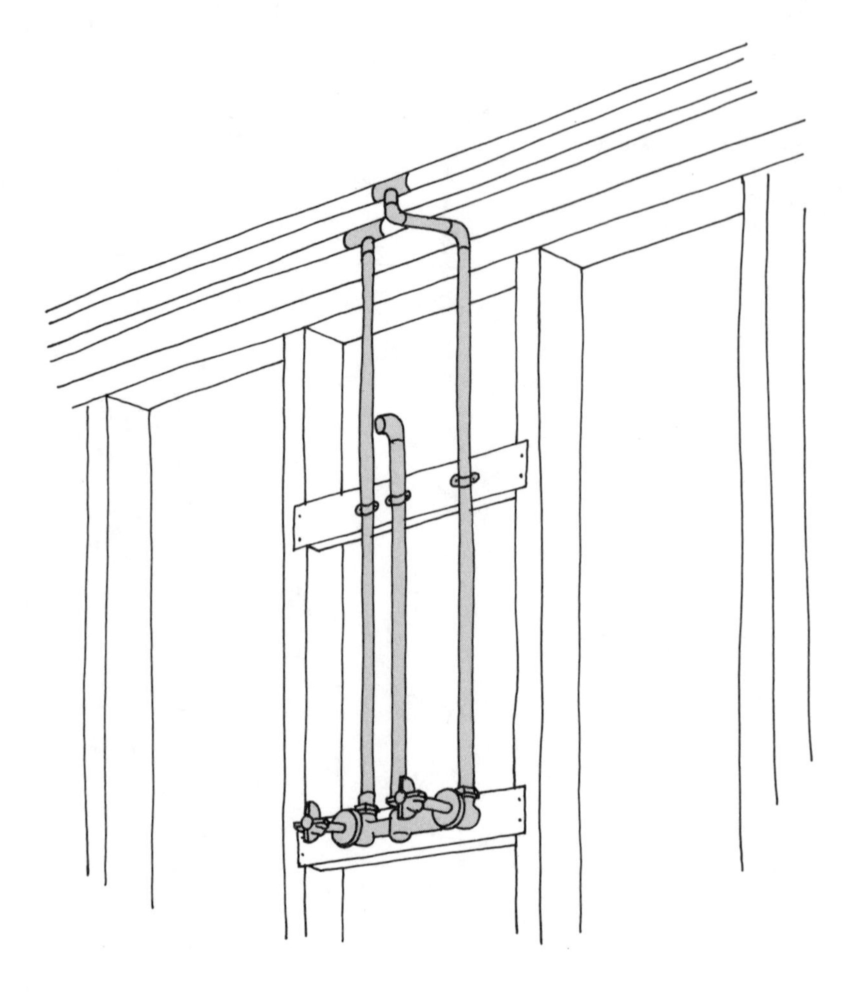

Basement shower *requires an existing floor drain. Tap into overhead supply pipes, build lines down wall. Use wooden braces and metal straps for support.*

ADDING A HOSE BIB

Installing an extra hose bib on an outside wall is a very simple job if there's a cold-water pipe running close to the point where the new connection is to be.

First drill a hole through the house wall so that its center is even with the center of the pipe. Tap into the supply pipe with a tee; this is a perfect place to use the saddle tee (see page 55). Screw on a nipple that will extend outside the wall about an inch. Then fit on the hose bib (check the bib when you buy it to see how much pipe should be extended outside the house).

If you live in a cold-winter area, you might want to install a freeze-proof hose bib. It's as easy to install as a conventional hose bib.

Add hose bib *by tapping into pipe that runs close to an outside wall. Galvanized run will require a union to reassemble.*

INSTALLING A BIDET

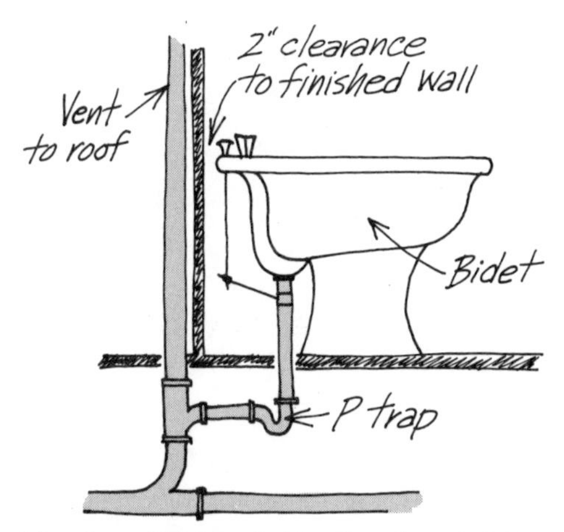

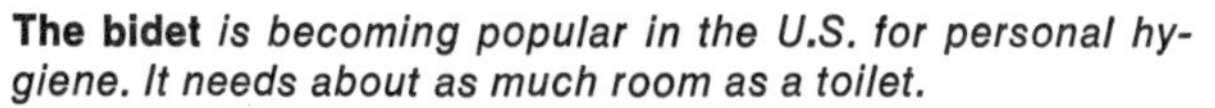

The bidet *is becoming popular in the U.S. for personal hygiene. It needs about as much room as a toilet.*

Plumbing for a bidet *is simpler than for a toilet; no closet bend is required, just a sink-type trap and drain.*

INSTANT RESULTS: A NEW SHOWER HEAD

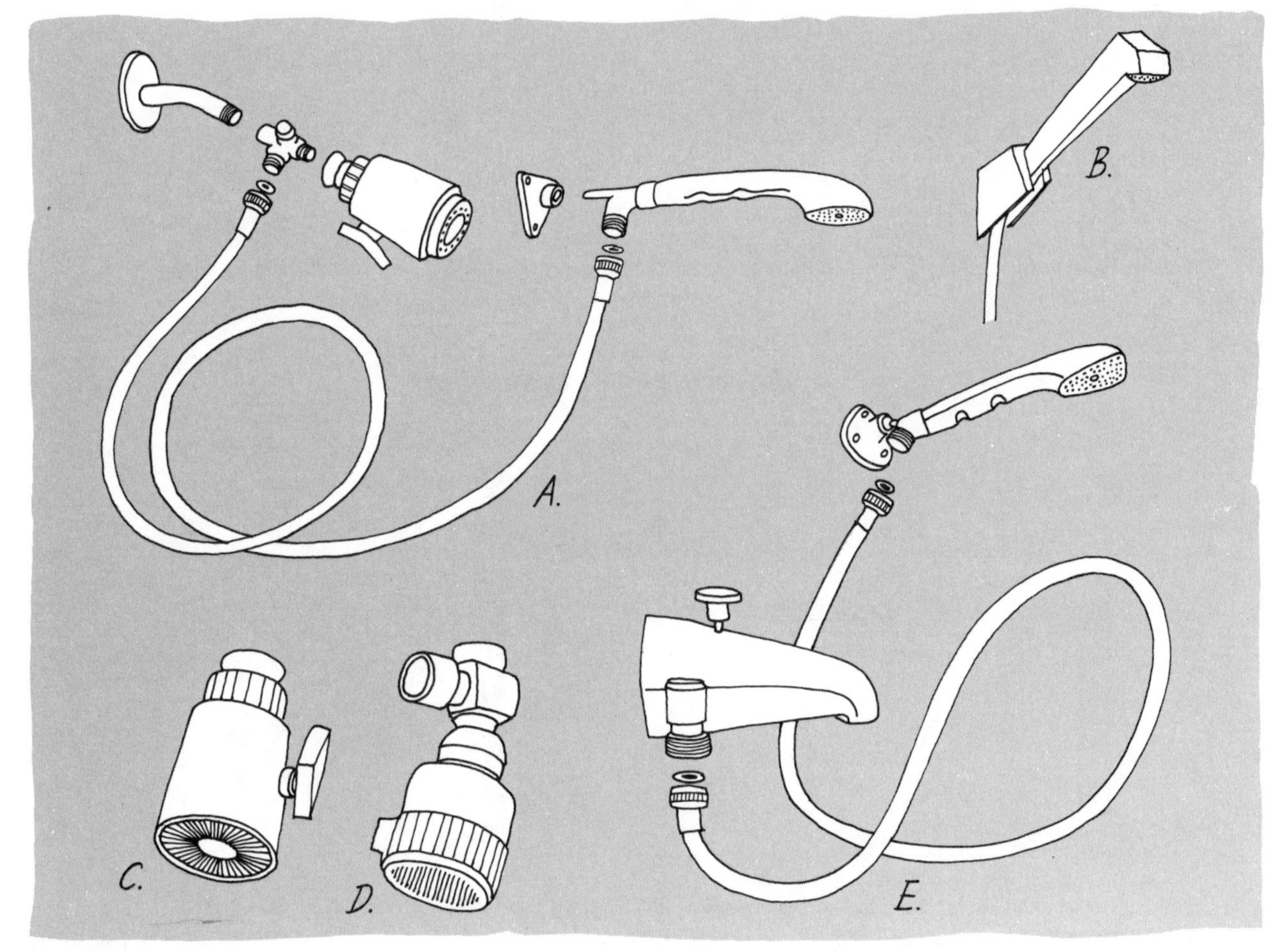

It's easy *to remove your old shower head and install a new one. The cable shower attachments (A, B, and E) attach to shower head or to a bath spout. C and D are two styles with a wide range of spray action.*

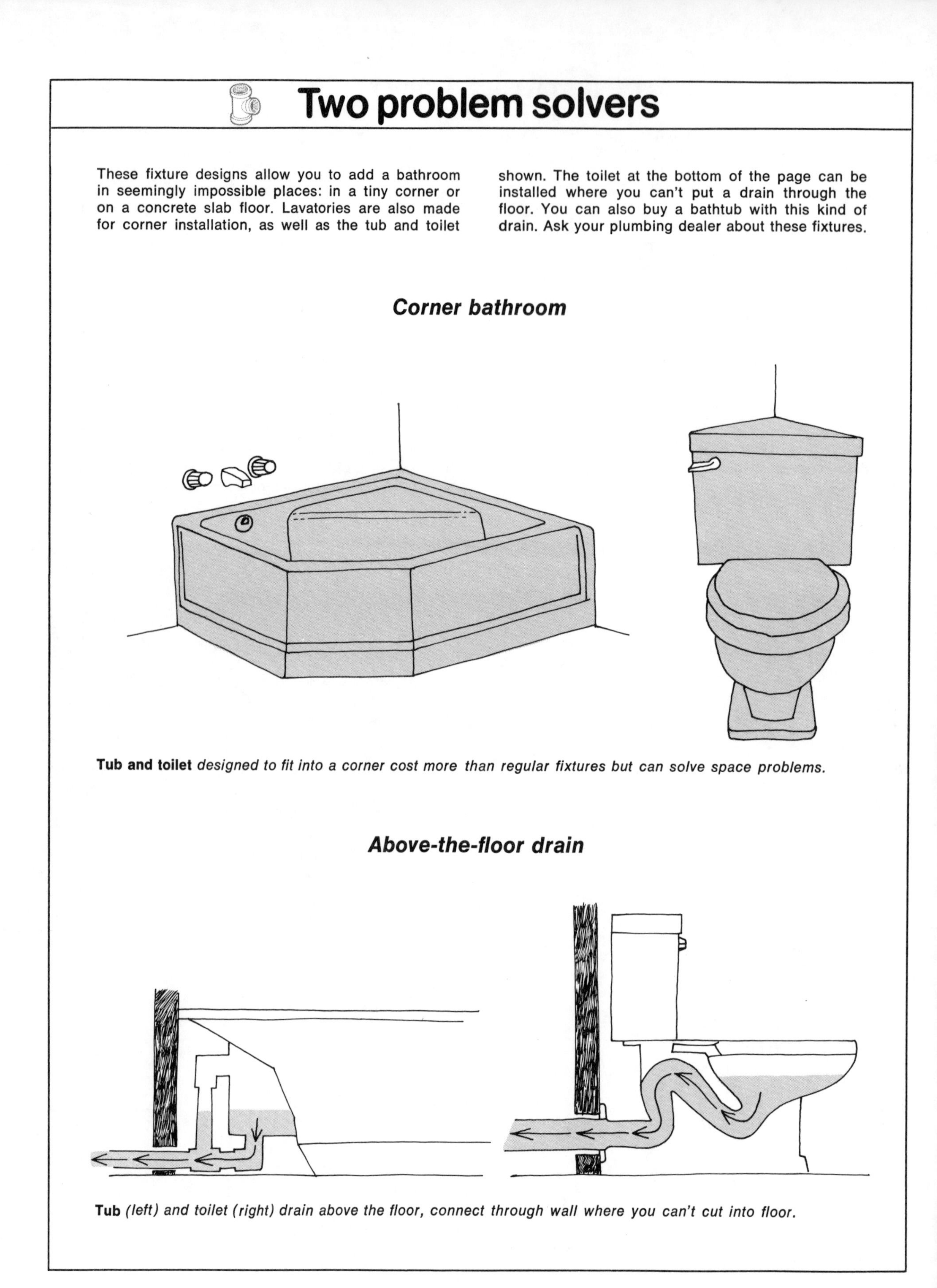

Two problem solvers

These fixture designs allow you to add a bathroom in seemingly impossible places: in a tiny corner or on a concrete slab floor. Lavatories are also made for corner installation, as well as the tub and toilet shown. The toilet at the bottom of the page can be installed where you can't put a drain through the floor. You can also buy a bathtub with this kind of drain. Ask your plumbing dealer about these fixtures.

Corner bathroom

Tub and toilet *designed to fit into a corner cost more than regular fixtures but can solve space problems.*

Above-the-floor drain

Tub *(left) and toilet (right) drain above the floor, connect through wall where you can't cut into floor.*

INSTALLING A SPRINKLER SYSTEM

Installation of an underground sprinkler system may sound like a forbidding task, but in fact the plumbing is simple, especially if you use plastic pipe. The hard work is in the digging.

Planning the system

Sprinkler manufacturers produce their wares in great variety. You can get bubblers for flower and vegetable beds and sprinklers that pop up when in use and sink back down so you won't trip over them. Sprinklers are available that water whole circles, half circles, quarter circles, and even squares. You can buy automatically timed controls and shrub sprayers. Decide what your needs are and then plan the installation carefully. Start by collecting literature from sprinkler manufacturers—the brochures will tell you the specifications of various sprinkler heads.

Determine the pressure of the water supplied to your house—ask your water service company or rent a pressure gauge and check the pressure yourself. To use the pressure gauge, fit the gauge onto the hose, making certain no other faucet is on in the house. Turn the water on full force and take a reading.

For a new sprinkler system, *you'll need to know water pressure. Screw gauge onto hose bib, turn water on full.*

You'll also need to know the sizes of your water meter and of the supply pipe you'll be tapping into. Your water company will give you meter size; see page 47 for a pipe size chart.

The next step is to make a sketch of the areas to be watered. Some equipment manufacturers provide graph paper to help you make a diagram to scale. Check the specifications of various sprinkler heads to find out how much area they cover. Mark the appropriate ones onto your scale drawing until the entire area you want to water can be covered.

The next step is to figure out which valves and pipes you will need. Sprinkler heads are normally arranged in groups, with all heads in a group controlled by one valve. Manufacturers suggest that you don't put two kinds of sprinklers—spray and bubble, for instance—in the same group. The valve size can't be bigger than the water meter and the service line.

The number of valves is determined by the number of sprinkler heads. Manufacturers will supply charts for their equipment that state how many heads each valve can control at various water pressures.

Group your valves as close together as possible, both for convenience and to save on costs. An anti-siphon valve will keep debris from going back into the water system; such a valve is required by some codes.

The size pipe you use to connect the supply main to control valves is determined by the distance between the two. The size pipe you need from valves to various sprinkler heads is determined by the volume of water that will flow through them. Manufacturers have charts available for their own products to help you determine this.

Hooking up the system

The way you tap into the water supply line will depend on your climate.

In cold-winter areas, connect with the water supply inside the house, just past the meter (see next page); then run the line back out through the wall to the control valves. Drill the

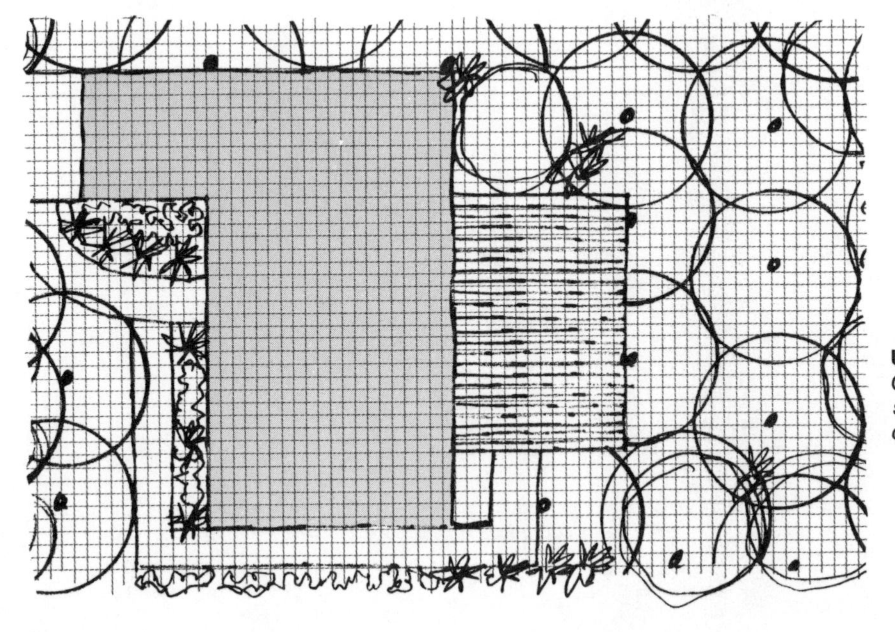

Use graph paper *to chart lawn area. Get manufacturers' specifications for sprinkler heads; then plot them on chart to cover desired areas.*

hole in the basement wall with a masonry drill. Note the shutoff valve and drain cock in the illustraton—you can shut off the system in winter and drain it at these points. (The pipe should be pitched back toward the drain.) Freezing isn't a serious problem when you use plastic pipe, for it will expand without bursting when the water freezes.

If you live in mild-winter areas and your water meter is outdoors, you can tap into the line outdoors and run a pipe underground to control valves. Install a shutoff valve for the whole system.

Under certain conditions you can connect the sprinkler system directly to an outdoor hose bib. Manufacturers' specifications will provide necessary details.

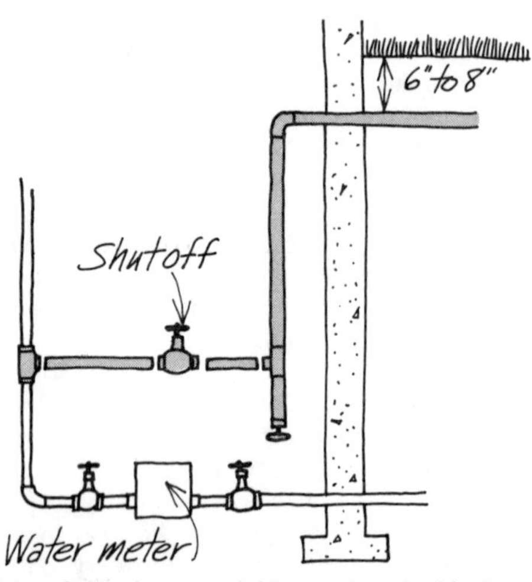

Cold winters? *Hook up sprinkler system inside house. To get outside, tap through basement wall with masonry drill.*

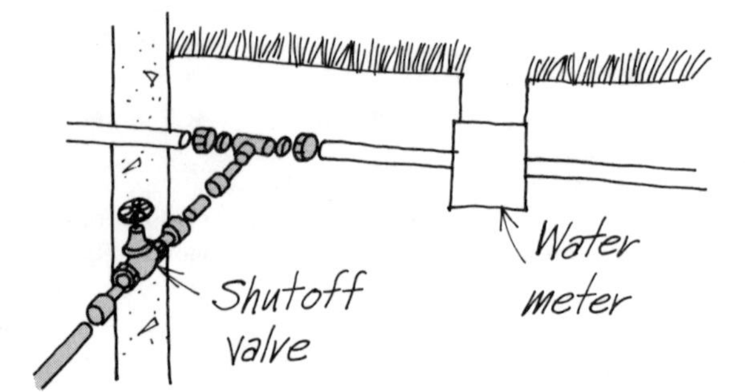

This outdoor watering system *taps into the water-supply pipe before it enters the house. Use this in mild-winter areas.*

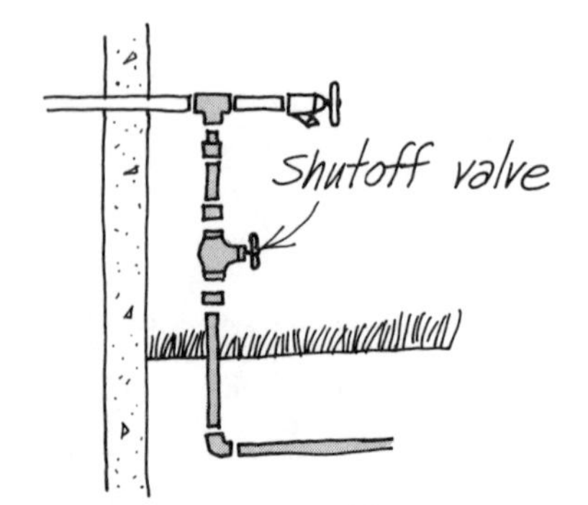

A simple way *to connect a new sprinkler system tap into an outdoor hose bib. Always include a shutoff valve.*

Assembling sprinkler pipes

It's a good idea to assemble the system above the ground to test it and adjust it before burying it. See pages 71-73 for instructions on assembling plastic pipe.

First assemble the groups of control valves. Hammer wooden stakes into the ground at each sprinkler head location. Lay out and assemble all circuits, including risers, but don't put on the sprinkler heads. Tie the risers to the stakes and turn on the water to flush out pipes and to check for leaks. Turn off the water, put on sprinkler heads, and turn the water on again. Adjust all sprinkler heads to cover the proper areas.

To bury the pipes, dig trenches 4 to 6 inches deep—you'll want the sprinkler heads to be flush with the top of the soil. Remember to slope the trenches so the pipes will drain

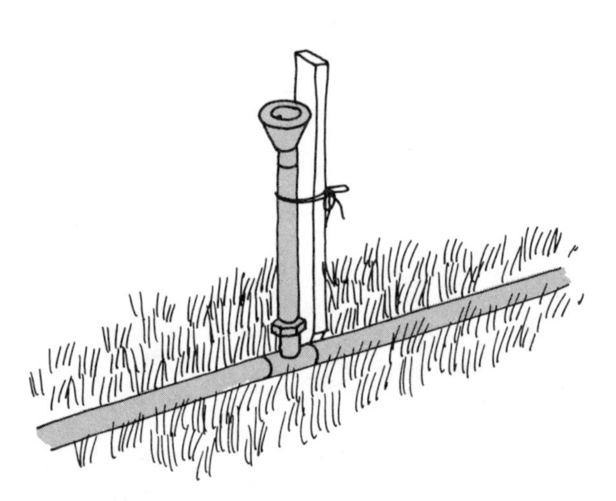

When assembling system *above-ground, tie risers to stakes to prevent movement that could damage pipes, fittings.*

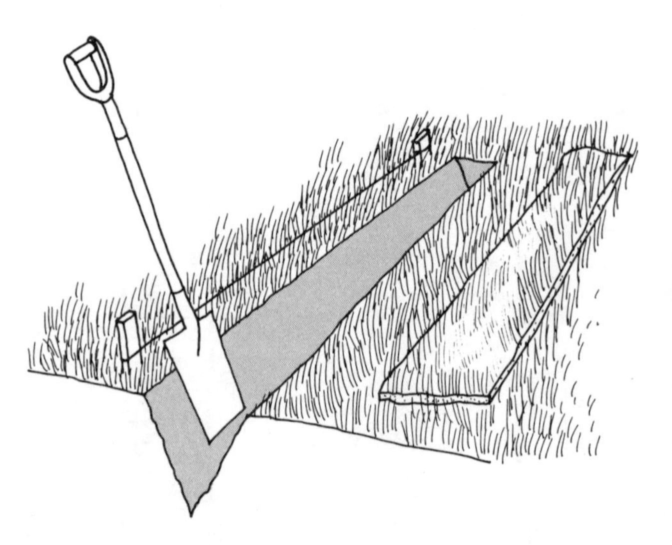

Bury sprinkler pipes *4 to 6 inches below the ground. To keep sod intact for reuse dig it with a flat spade.*

back toward the drain valve. If you're digging in a lawn, dig with a flat spade to keep the sod intact.

To burrow under a sidewalk or driveway, attach a piece of threaded galvanized pipe to a garden hose (you can get an adapter for this purpose). Dig an access hole beside the walk or driveway and put the hose with pipe attachment into the hole. Turning on the water full force, use the pipe to burrow a hole under the surface.

Right. *To dig under sidewalk or driveway, attach a length of pipe to hose, force into ground, turn water on full blast.*

Outdoor repairs

When the outdoor watering system isn't working, the problem is usually a clogged sprinkler head or a leaky hose. Both are eay to fix.

Unclogging a sprinkler head

Sprinkler heads are quite simple mechanically. If they misbehave, it probably means the water jets are clogged with dirt. The illustration below shows you how to clean the common type of sprinkler head shown below (many are similar to this one) by taking the following steps:

1. Insert a screwdriver in the notch between the core and the housing and lift the core.
2. Use a wrench to unscrew the top piece of the core.
3. Dislodge any foreign matter in the core with a wire or other sharp object. (You may need to tighten or loosen the deflector.)
4. With the top part of the core still off, turn on the sprinkler to flush dirt out of the pipe and the lower part of the unit.
5. Replace the top part of the core unit after flushing.

If your sprinkler head doesn't have a core that extends out where you can unscrew it, loosen the screw in the top of the head several turns, and then turn the water on to flush out the dirt. Let the water run for several minutes; then tighten the cap down again and check the action of the water.

If a sprinkler head in a system consistently malfunctions, it should be replaced. A special wrench will loosen it from the tee that secures it to the main line. This wrench is designed to avoid uneven strain on pipes or fittings.

Patching a leaky hose

Hoses with pinhole leaks can be fixed with a wrapping of electrician's tape—the black, shiny-surface, plastic type. The hose should be completely dry, and hose and tape will bond more firmly if both are warm.

Wrap the tape at least 2 inches on each side of the leak, overlapping each turn by half the width of the tape. Three or four layers should be enough.

If hose is split but worth saving, cut out the damaged section and join the two remaining sections with a special clamp, shown below. (Cut the hose with a knife.) The claws of the clamp can be tapped into place with a hammer (and pried up again with a screwdriver for reuse).

When a hose end splits, you can install a new coupling in much the same way. Cut off the old coupling and two inches of hose; then clamp the new coupling in place.

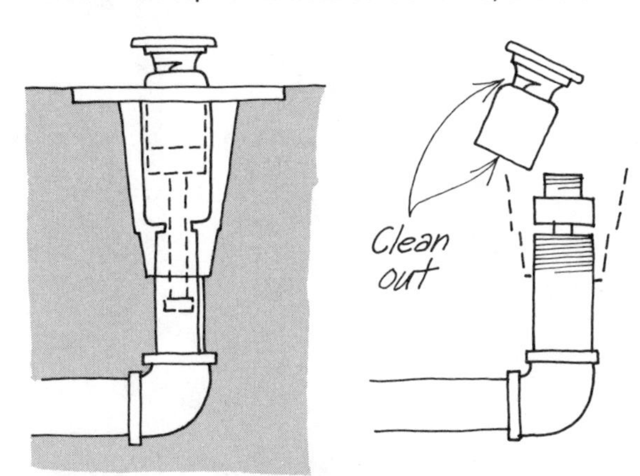

Sprinkler heads *often become clogged. Most unscrew easily. Once removed, flush them out with water.*

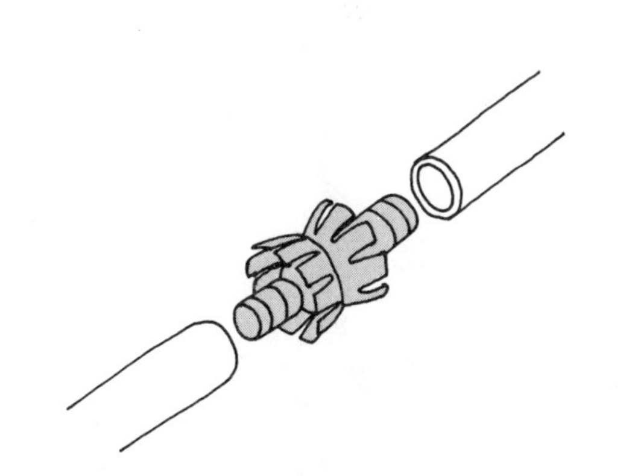

To repair leaking hose: *cut out leaking section, rejoin hose with clamp, hammer claws down.*

Pipefitting know-how

Plumbing pipe is made of all sorts of materials—copper, iron, steel, stainless steel, brass, and plastic. Cement pipe and vitreous clay pipe are used in sewers. Glass pipe is used where acids and other strong chemicals are drained away. Plastic pipe, originally made for outdoor sprinkler systems, is now used indoors as well (code permitting) for water supply, drainage, and vents.

Pipefitting involves just a few basic steps, but these do take some mechanical skill and care to execute without a hitch. You'll find the most difficulty working with heavy cast-iron and steel pipe. Read the instructions in this chapter carefully before you start to work.

The two categories of pipe you will work with are supply pipe and DWV (drain-waste-vent) pipe.

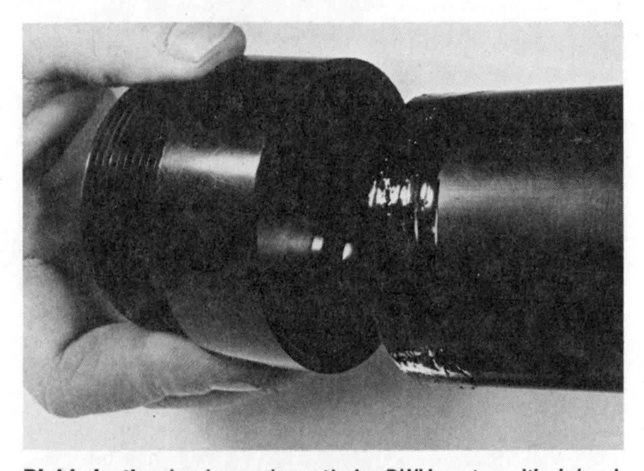

Rigid plastic *pipe is used mostly for DWV system. It's joined with solvent cement—here to a threaded adapter.*

Galvanized steel *pipe is used extensively in supply systems. Pipe and fittings are malleable iron and are threaded.*

Rigid copper *pipe can be used for both supply and DWV. (DWV sizes are expensive.) These supply pipes are soldered.*

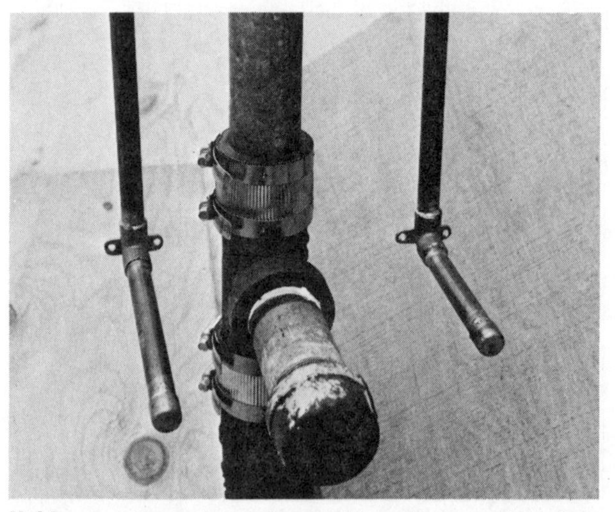

Hubless cast-iron *pipe makes working with large DWV sizes much easier. Pipe ends fit flush and are clamped together.*

SUPPLY PIPE

Supply pipes in homes are most often galvanized iron or steel, copper, or plastic. They are usually not larger than 1-inch nominal inside diameter. (The actual diameter is seldom the same as the nominal diameter, but it is the nominal diameter that you will use when selecting pipes and fittings.)

When you've finished fitting the supply pipes, you'll want to test the new joints for leaks. Some codes require that your work be checked by an inspector; codes may prescribe the method for testing for leaks.

Turn the water on and examine each joint for escaping water. If you've built an entire supply system, you can check for leaks before water enters it, using an air compressor. To do this, make certain all outlets are off, open one to pump in the air, and then shut it off. Check at all the joints for the sound of air escaping. If you don't hear this sound, the system is tight.

Galvanized pipe

Galvanized pipe is iron or steel pipe that has been treated with zinc to retard corrosion. Despite this, of the three most common kinds of pipe in use in home water supply systems, galvanized is the most susceptible to corrosion. Copper is virtually noncorrosive; plastic doesn't corrode at all. Galvanized pipe can also accumulate calcium deposits.

For many years galvanized pipe was almost the only kind used for house water supply systems, and it is still used very widely today (it's usually cheaper than copper and is much stronger than either copper or plastic). If you have an older house, the chances are good that you have galvanized pipe in your supply system.

For the home plumber the chief disadvantage of galvanized pipe is its weight, which makes it more difficult to work with than either copper or plastic.

How to measure a run

Before you cut any pipe, your measurements must be exact. Rigid pipes won't "give" to compensate if they're too long, and no one has invented a pipe stretcher to take care of the too-short lengths. The first tool that should go into your pipefitter's kit is a measuring stick or carpenter's folding rule. Use a rigid ruler for measurements that must be precise.

Fittings are part of a pipe run; take their dimensions into account when measuring. First, measure the entire run from the center of the fittings on each end. Next, you need to account for the difference between that center-to-center distance and the distance between the pipe end and the center of the fitting. To do this for galvanized pipe, subtract a distance identical to the nominal inside diameter of the pipe for each fitting. For ½-inch pipe, subtract half an inch; for 1-inch pipe, subtract 1 inch. So if you have a run of ½-inch pipe 4 feet long with a fitting on each end, you'll need pipe that measures 3 feet 11 inches long.

This calculation won't necessarily work for pipe other than galvanized. For copper and plastic, establish the face-to-face distance (the face is the outside edge of the fitting) by measuring it directly or by first measuring the center-to-center distance and then subtracting the center-to-face distance for each fitting. Then measure from the inside shoulder (where the pipe stops) to the outside edge of the fitting, adding that distance to the face-to-face distance between the two fittings.

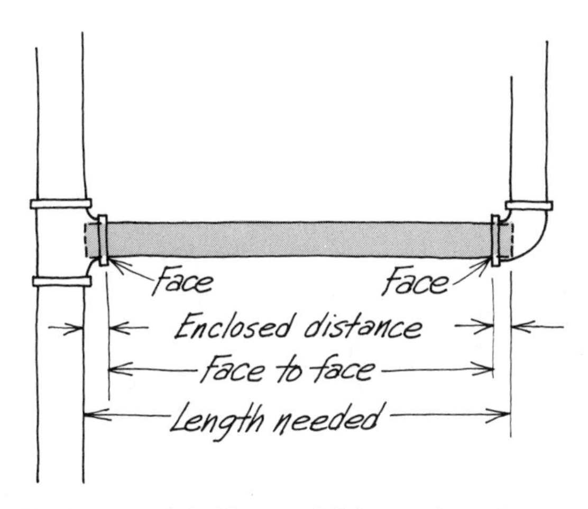

Pipe length *needed is sum of 1) face-to-face distance between fittings and 2) distance pipe goes into fittings.*

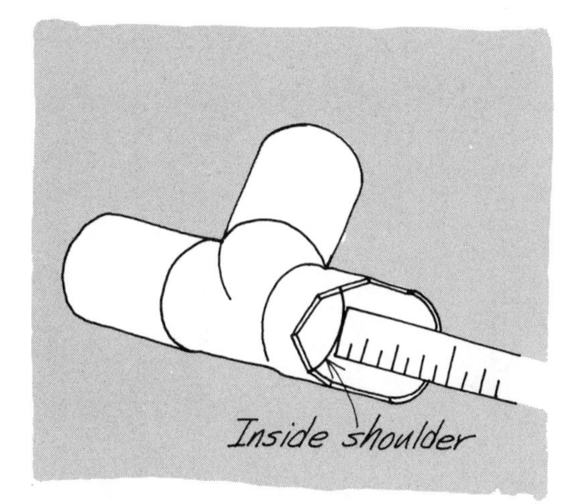

Copper and plastic: *To find distance pipe goes into fitting, measure from inside shoulder of fitting to face.*

Cutting galvanized pipe. Before you cut any pipe, make sure you have measured and marked the correct length exactly (see page 63). Many plumbing supply stores will cut and thread lengths of galvanized pipe for you. If you're going to do it yourself, you'll need some special equipment: a pipe cutter or a hacksaw with a 24 or 32-tooth-per-inch blade, a pipe vise, a reamer, and a file.

It's important to cut perfectly straight so the pipe won't leak around the fitting. The easiest way to insure a square cut is to use a pipe cutter, a tool that fits onto the pipe end exactly and cuts true.

After the pipe is cut, you'll find some burrs both inside and outside. File off the outside burrs; use the tapered reamer to grind off the inside ones (left in, they would restrict the flow of water). Clean the pipe end inside and out with a rag.

Threading galvanized pipe. The threading die should have the same nominal diameter as the pipe—for example, a ½-inch die for a ½-inch pipe. Fit the die into the stock (the handle). Then slip the die over the end of the pipe. Exerting force inward toward the pipe, begin rotating the handle clockwise. When the die bites into the pipe, you can

Cutting, threading, and installing galvanized pipe

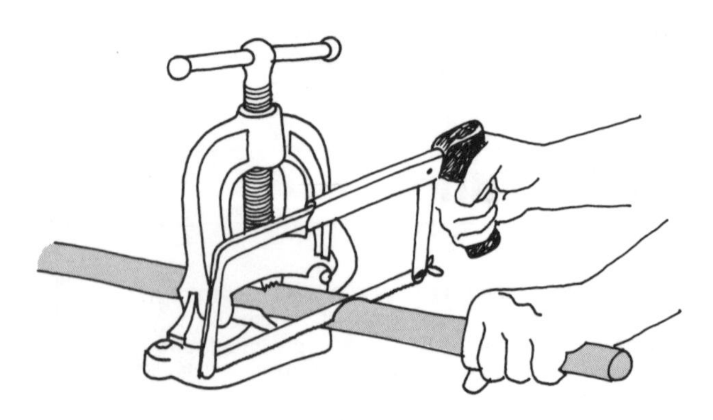

Step 1. *Fit pipe into the vise. (Some have pipe jaws under regular jaws.) Cut with hacksaw or a pipe cutter.*

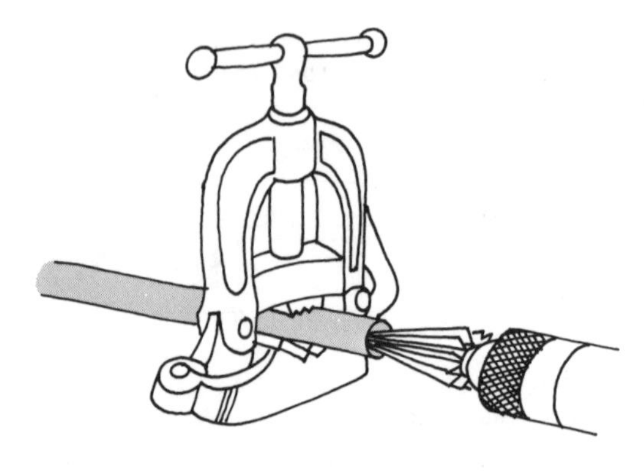

Step 3. *File off burrs on the inside of the pipe with a reamer. Vise holds pipe steady during the reaming process.*

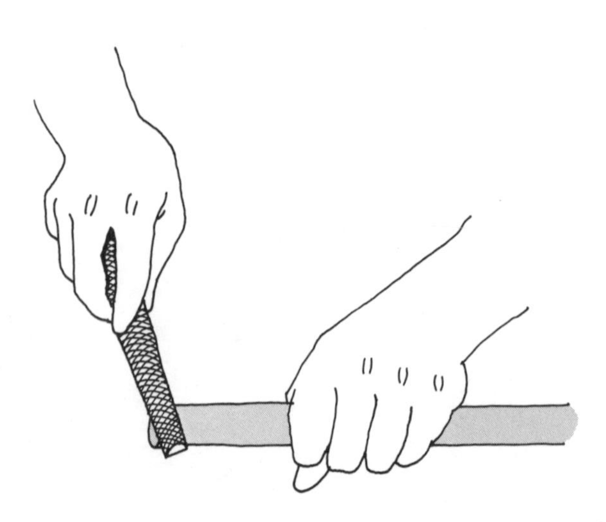

Step 2. *Cutting will leave burrs on both the inside and outside of the pipe. File off the outside burrs.*

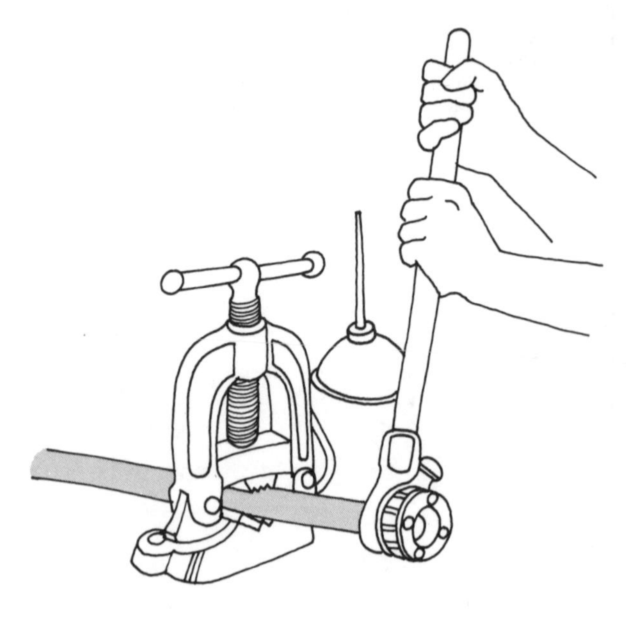

Step 4. *Fit pipe threader over pipe end and tighten it. Exert force at first while turning to start cutting action.*

stop pushing and simply continue the clockwise rotation. (Once begun, the threads will pull the die along without your needing to exert any force.) Apply generous amounts of cutting oil as you turn the threader. If the threader gets stuck, you probably have some metal chips in the way; back the tool off very slightly and blow the chips off. Continue threading until the pipe extends about one thread beyond the end of the threader. Finally, remove the threader and clean off the threads with a stiff wire brush.

Installing galvanized pipe. Once threaded, the pipe is ready to be installed with its fittings (galvanized pipe requires malleable iron fittings). Coat the threads of the pipe with pipe joint compound or wrap them with fluorocarbon tape. Do not coat the inside of the fitting. Screw the pipe and fitting together by hand as far as you can. Do this slowly—done too fast, joining creates heat which causes the pipe to expand; later, the pipe shrinks and the joint becomes loose.

After you've tightened the pipe and fitting by hand as far as you can, finish with two pipe wrenches. Use one on the fitting and one on the pipe, turning in opposite directions. (Always turn pipe wrenches toward the opening of their jaws. See drawing below.)

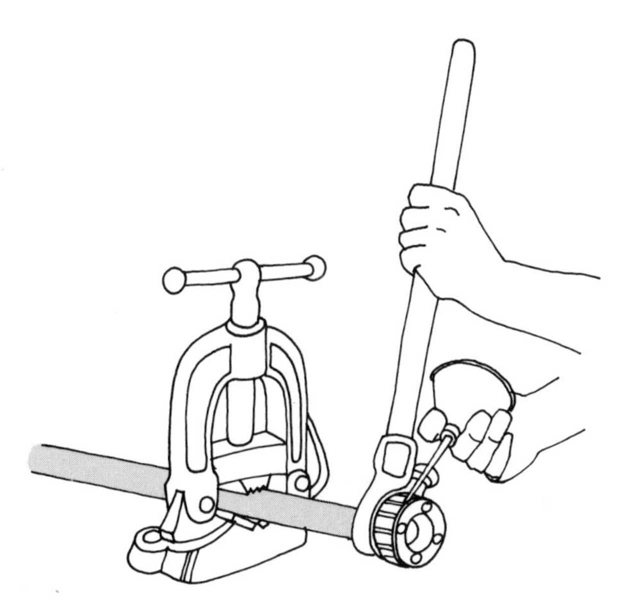

Step 5. *While cutting threads, use generous amounts of oil. Chips may stop threader. Back off, brush chips away.*

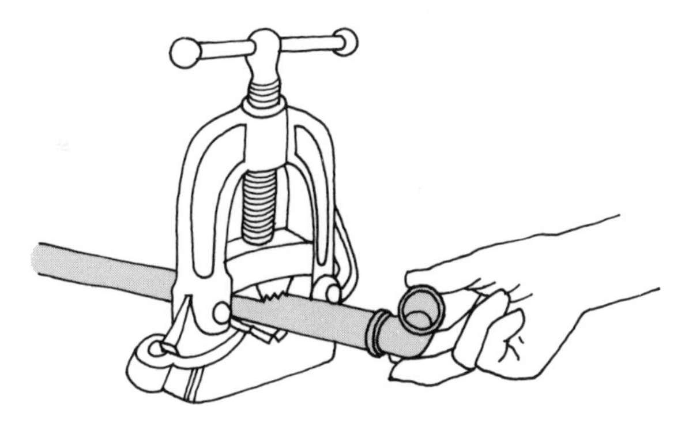

Step 7. *Clean out the threaded inside of fitting; then screw it onto the pipe end by hand as far as it will go.*

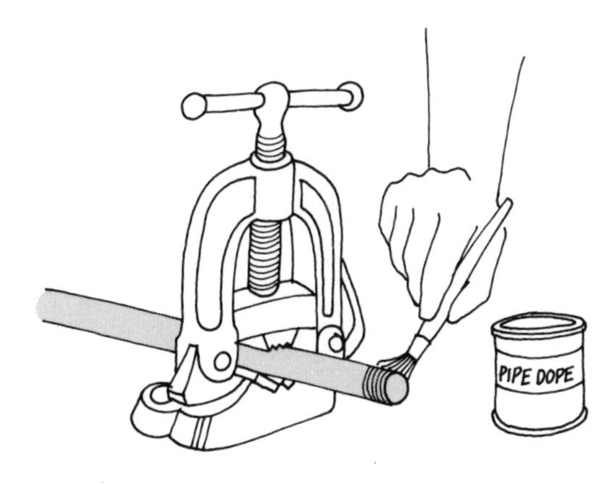

Step 6. *Threads cut, pipe end cleaned off, apply pipe joint compound to threads to help seal the completed joint.*

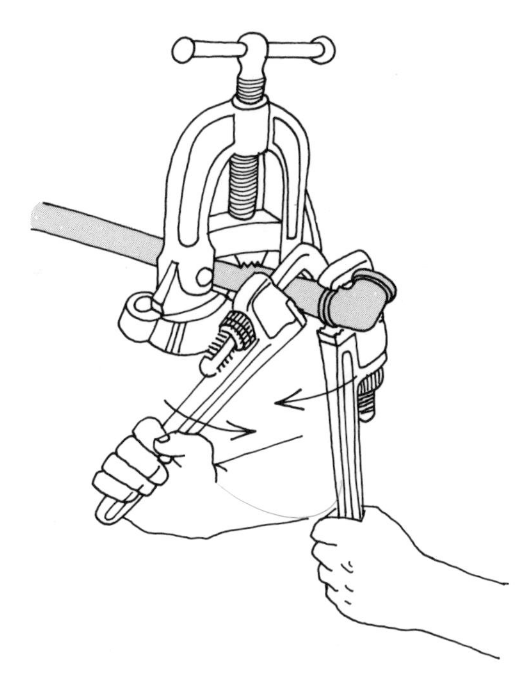

Step 8. *Use two wrenches to tighten joint—one on pipe, one on fitting. Turn slowly in opposite directions.*

Which fitting for which pipe?

Fittings are used to join sections of pipe, to change the direction of a pipe run, to reduce or enlarge the pipe end so that it can be joined to a different size pipe, and to plug the end of a pipe. So that you know exactly which fittings you need, consult this chart before you go to the plumbing supply store. Specify the following information for each fitting:

Material. The most common fittings used for galvanized, copper, and plastic pipe are shown. Some fittings allow you to attach pipe made of one material to pipe made of another.

Threaded or smooth. Specify whether you want to buy threaded fittings (such as the malleable iron fittings used for galvanized pipe) or unthreaded fittings (such as copper or brass fittings used for soldered joints on copper pipe and almost all plastic fittings for plastic pipe). Male fittings are threaded on the outside; female fittings on the inside.

Size. Specify the size pipe you are joining. When ordering reducing fittings, give the largest size first, then the smallest. For Ts, Ys, elbows, and crosses, give the main run first and then the outlet.

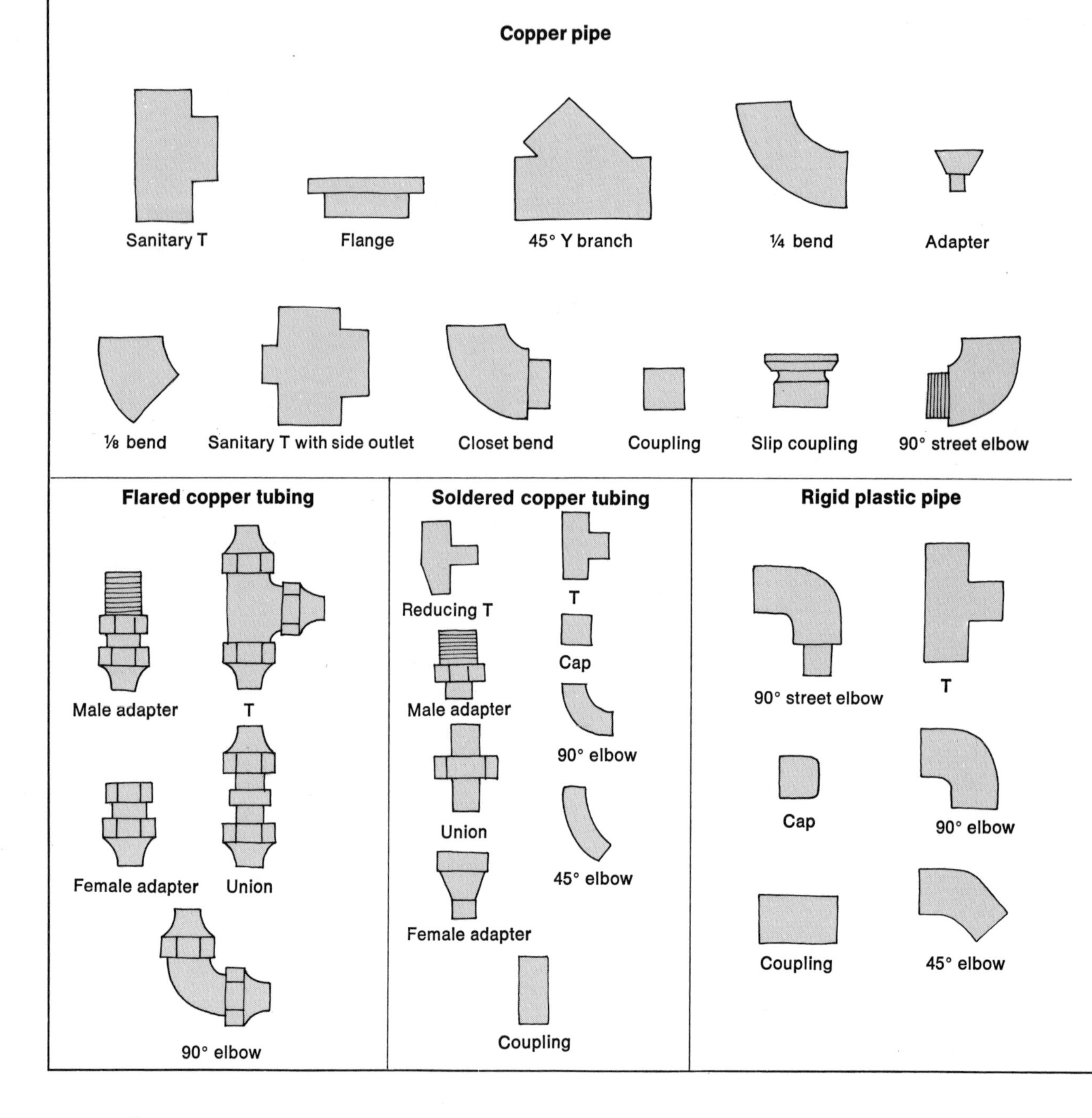

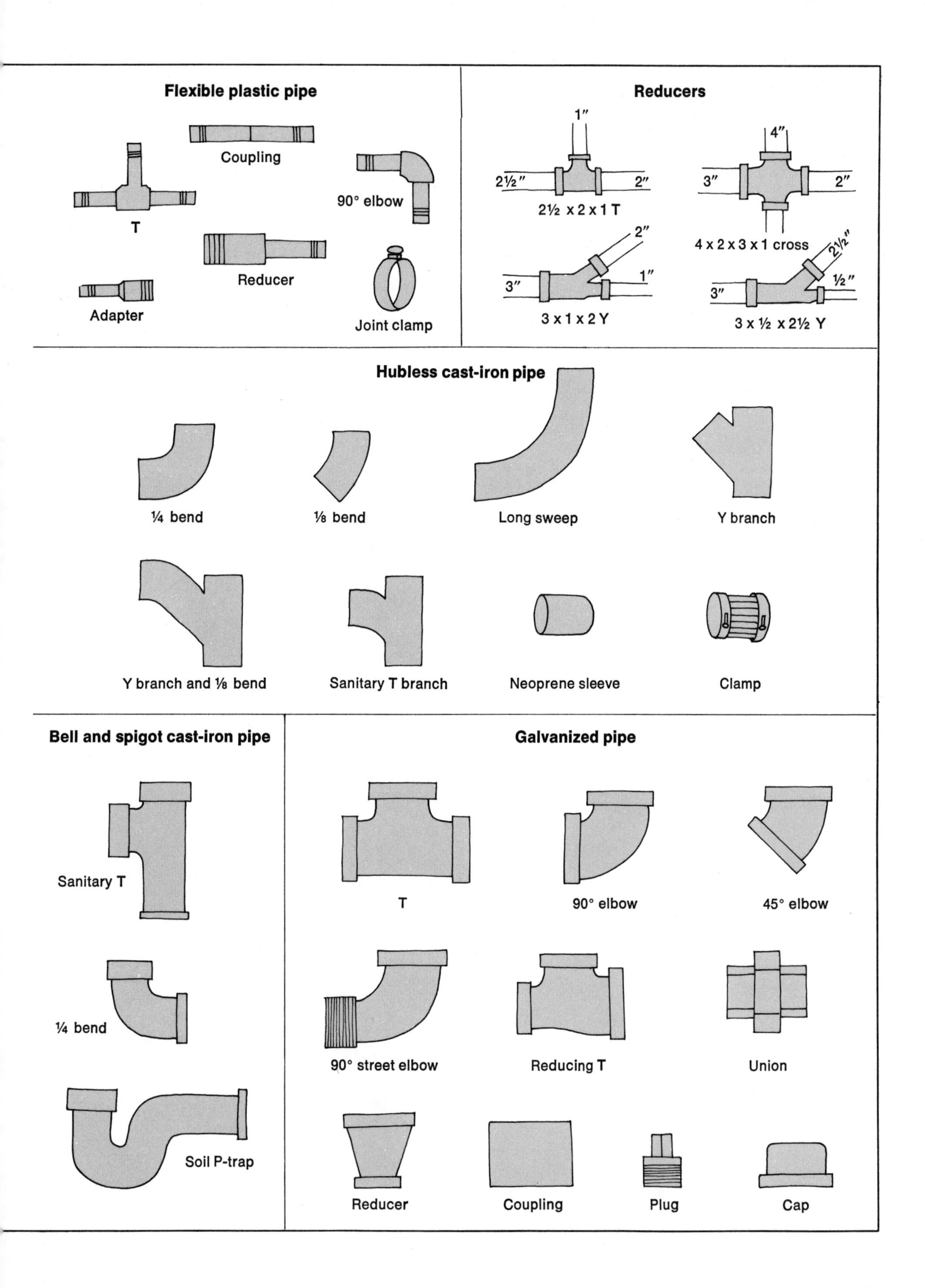
Flexible plastic pipe
Coupling
T
90° elbow
Reducer
Adapter
Joint clamp
Reducers
1″
2½″
2″
2½ x 2 x 1 T
4″
3″
2″
4 x 2 x 3 x 1 cross
2″
3″
1″
3 x 1 x 2 Y
2½″
3″
½″
3 x ½ x 2½ Y
Hubless cast-iron pipe
¼ bend
⅛ bend
Long sweep
Y branch
Y branch and ⅛ bend
Sanitary T branch
Neoprene sleeve
Clamp
Bell and spigot cast-iron pipe
Sanitary T
¼ bend
Soil P-trap
Galvanized pipe
T
90° elbow
45° elbow
90° street elbow
Reducing T
Union
Reducer
Coupling
Plug
Cap

Copper pipe

Copper pipe is very popular with both amateur and professional plumbers. It's highly resistant to corrosion, very lightweight, easy to join, and rugged enough for home use. Copper pipe is tougher than plastic pipe but less strong than galvanized pipe. It's usually more expensive than either the plastic or galvanized types.

Because copper is a soft metal, you'll want to be careful not to damage it when working with it. Use electrician's tape to cover the jaws of wrenches and vises that have teeth. A special wrench called a "strap wrench" will not damage the pipe.

Copper comes as rigid pipe or as soft tubing. The rigid pipe has three thicknesses: K, the thickest wall; L, medium wall; and M, thin wall. Though M should be adequate for home use, check your local plumbing code to make sure it's acceptable.

Tubing comes in two thicknesses: K is heavy-duty; L is medium weight and is good for most home use.

Rigid copper pipe. You can use threaded joints on types K and L, but because of its thin walls, type M will take only soldered joints (also called "sweat joints"). Soldering is a simple process, though, and works very well on K and L, too. And soldered fittings for copper pipe are far less expensive than threaded ones.

• *Cutting copper.* Measure the copper pipe, accounting for the distance taken up by the fittings in the run (see page 63). You can cut it with either a hacksaw (24 or 32-tooth-per-inch blade) or a pipe cutter, but using the pipe cutter is easier. Use a pipe cutter designed for copper pipe (see photo opposite); the procedure is the same as for cutting galvanized pipe.

If you use a hacksaw, the best way to insure a straight cut is to hold the pipe in a miter box.

Since copper is much softer than iron and steel pipes, it can get crushed in a vise. If you do use a vise, clamp the pipe several inches away from the end so there's no chance of damage where the fitting will go.

After you've made the cut, clean off the burrs inside with a half-round file. A cutting tool will leave larger burrs on copper pipe than a hacksaw will. To minimize burrs, get a cutter with a thin blade.

• *Soldering copper.* To solder a joint, you'll need a small torch (propane or butane torches work well for this), some 00 steel wool or very fine sandpaper or emery cloth, a can of soldering flux, and some solid core wire solder (50 percent tin, 50 percent lead is best).

Use the steel wool, sandpaper, or emery cloth to polish the outside end of the pipe until it is shiny. Clean the pipe end inside and out; clean the inside of the fitting. With a small, stiff brush, apply flux around the inside of the fitting and around the outside of the pipe end. Flux will prevent the copper from oxidizing when it's heated.

Place the fitting on the end of the pipe (if the fitting is fixed to another pipe, place the pipe in the fitting). Turn the pipe or the fitting back and forth once or twice to spread the flux evenly. Next, position the fitting correctly and heat it with a torch. It's important not to get the fitting too hot because the flux will burn when overheated. Test heat level this way: the joint is hot enough when solder will melt on contact. Touch the wire to the joint once in a while as you're heating. The minute the wire melts, the joint is ready.

Take the torch away and touch the solder to the edge of the fitting; capillary action pulls solder in between the fitting and the pipe. Keep soldering until a line of molten solder shows all the way around the fitting. With a rag, wipe off the surplus solder before it solidifies. Be careful not to bump or move the newly soldered joint the least bit because any jarring will displace the moist solder and cause leaks.

Cutting and soldering copper pipe

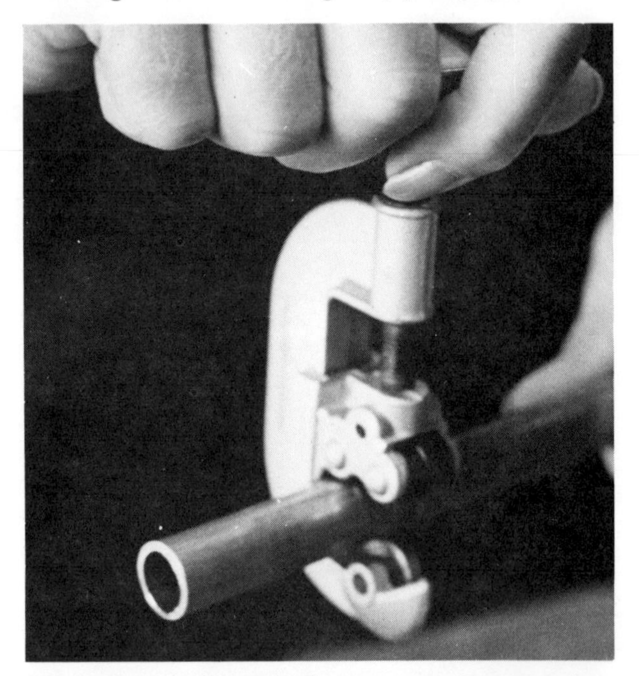

Step 1. *Special pipe cutter insures a good square cut on copper pipe. (With hacksaw, put pipe in miter box.)*

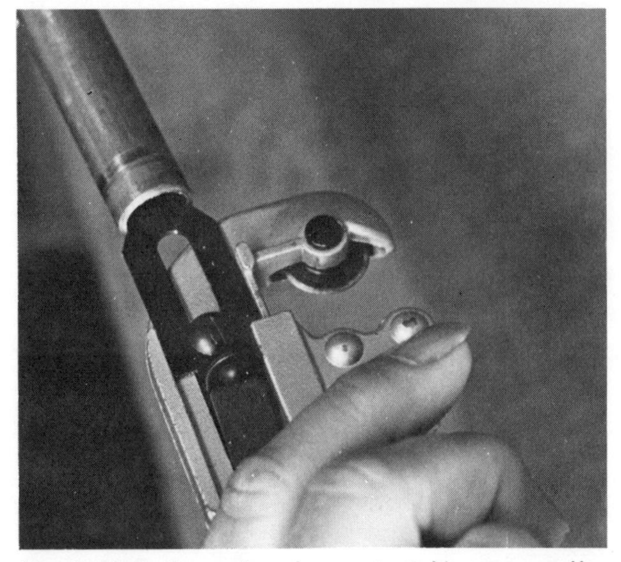

Step 2. *Most pipe cutters have retractable reamers. Use reamer to ream out burrs. Sand off outside burrs.*

Step 3. *Using 00 steel wool on the end of the pipe, bring it to a high polish. Wipe off any bits of wool with dry cloth.*

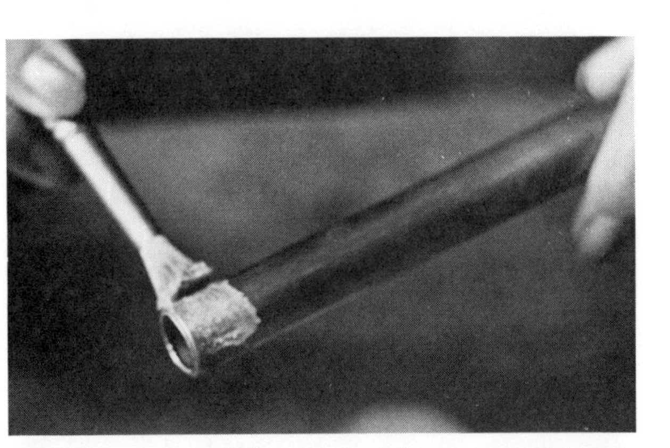

Step 4. *You can apply flux with your fingers or a stiff brush. Apply thick coat to outside of pipe end, thin coat inside fitting.*

Step 5. *Apply the torch flame to the fitting, not the pipe. Test occasionally—when solder melts, fitting's hot enough.*

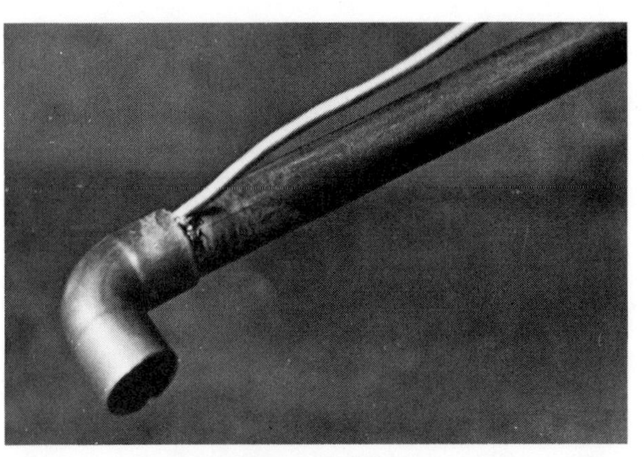

Step 6. *Apply solder at fitting. Use diameter of pipe to measure solder: 1 inch of solder for 1 inch pipe, etc.*

Step 7. *Soldering done, quickly wipe smooth the bead of solder at edge of fitting. These are called "sweated joints."*

Copper tubing. A good reason to use flexible copper tubing is that you can bend it, frequently eliminating the necessity for joints. It's often a good choice for remodeling jobs because it will fit into places where you might otherwise have to tear out a wall to install rigid pipe.

You can join tubing either by soldering or by using flare or compression joints. Although these are more expensive than soldered joints, assembling them is nearly as easy as splicing a garden hose, and they're simple to take apart again for repairs on the system.

Because copper tubing will sag easily, make certain it is well supported with metal straps or pipe hangers, especially if you're using it for a horizontal run.

Cutting and soldering copper tubing. For home use, type L (medium wall thickness) is a good choice. You can cut it either with a hacksaw (24 or 32-tooth blade) or a tubing cutter, which is very effective and easy to use. The procedure is the same as for cutting rigid copper pipe and galvanized pipe. If you use a hacksaw, hold the pipe in a miter box to insure a good square cut.

After the cut is made, ream out the burrs with a half-round file and smooth and clean off the outside with 00 steel wool or very fine sandpaper or emery cloth. You may discover that the end is out of round—this sometimes happens to copper tubing when it's uncoiled. You can either cut off the portion that is lopsided or use a sizing tool to make it round again.

To solder joints, follow the same procedure as for rigid copper pipe.

Flare and compression fittings. Instead of soldered fittings on copper tubing, you can use flare and compression fittings. These are very easy to install and equally easy to take apart. They're expensive, though.

To assemble a flared joint you'll need a flaring tool. One type simply fits into the end of the tube and is pounded with a hammer. A screw-down type clamps onto the pipe, and the tapered point screws down to produce the flare.

Compression fittings are made without the flare. First, a flange nut fits over the end of the tube. Then a compression ring slides on. Next, the body of the fitting is placed against the end of the tube, and the flange nut is screwed onto it, compressing the ring very tightly between fitting and pipe.

Assembling a flare fitting

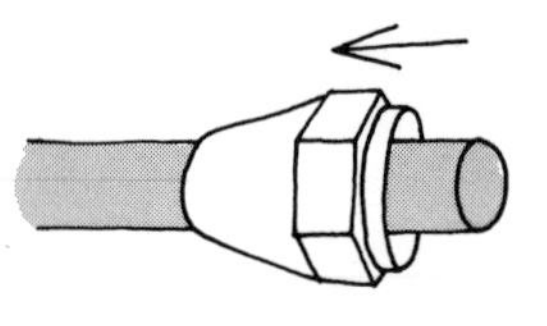

Flare fittings *for copper pipe are expensive but easy to assemble. First fit threaded fitting over pipe end.*

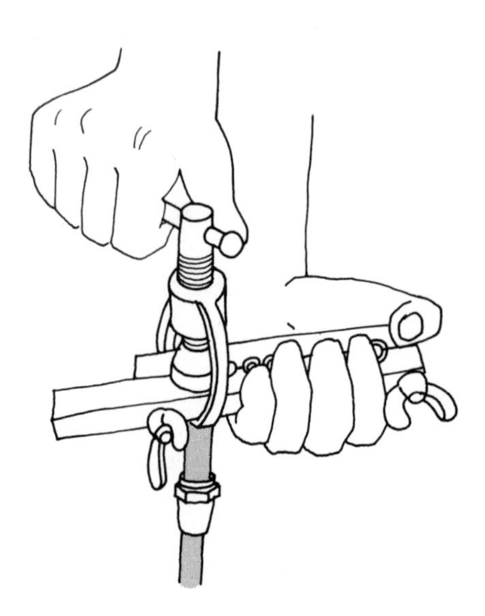

Special flaring tool *does precise job. Clamp it down over pipe end and screw tapered head into pipe, creating flare.*

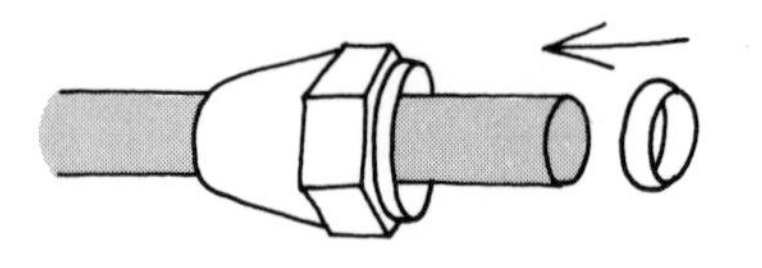

After the threaded fitting *is in place, slide the small metal gasket over pipe end next to the fitting.*

Once flare is made, *the end of this threaded fitting will fit snugly into it. Slip female end over it, screw down tight.*

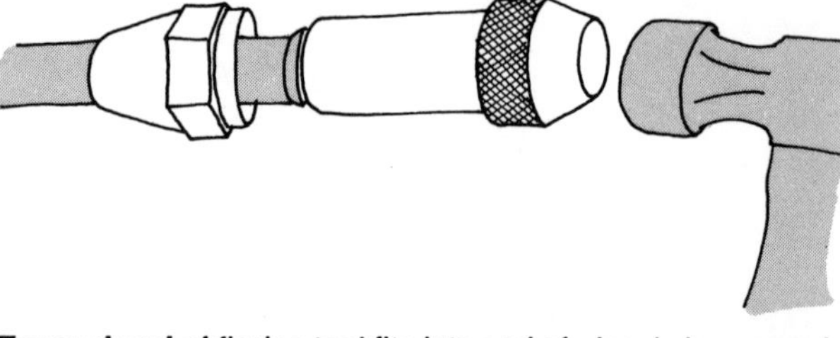

Tapered end *of flaring tool fits into end of pipe, is hammered in to make flare. Use vise for support.*

Assembling a compression fitting

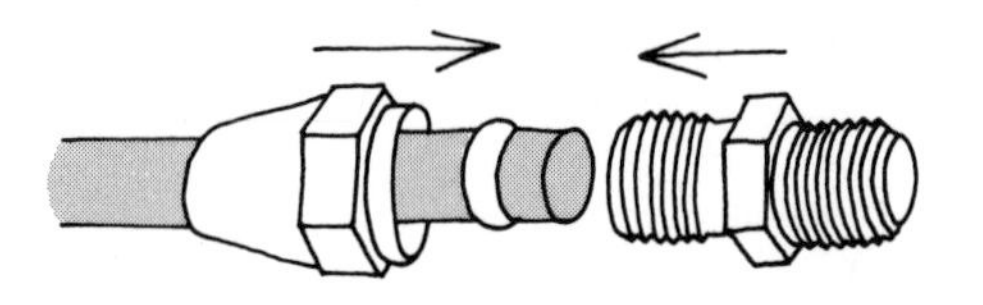

Compression fittings *use threaded male and female connections and compression ring, don't require initial flaring.*

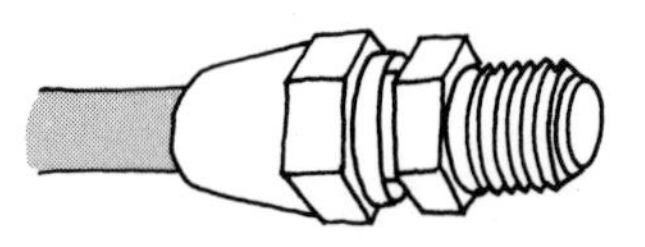

Once joined, *flare and compression fittings look the same. Disassembling is a simple task—unscrew the female ends.*

Bending copper tubing. Though you don't need any special tools to bend copper tubing, you must be careful to keep the pipe walls from caving in. With your thumbs together at the middle of the bend, pull the ends in toward the center with your fingers little by little, moving your thumbs apart a fraction of an inch at a time. Continue until the bend is complete.

For very sharp bends on short lengths of tubing, try filling the tube with sand for internal support while you are bending it.

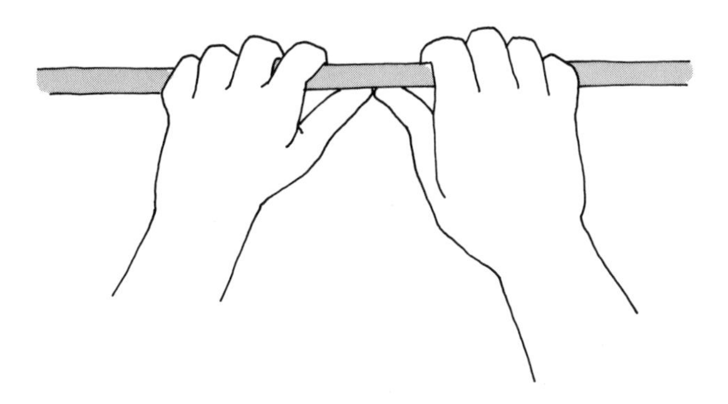

Bend copper tubing *carefully to prevent crimping. Begin with your thumbs together at the middle of the bend.*

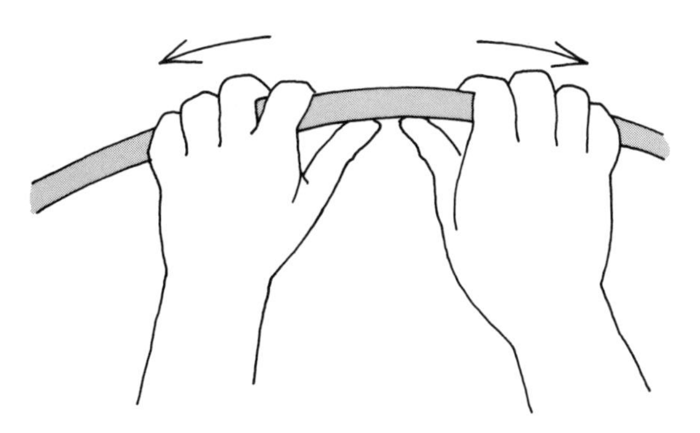

Pull the ends in *toward the center of the bend, exerting pressure with your fingers. Move thumbs slowly apart.*

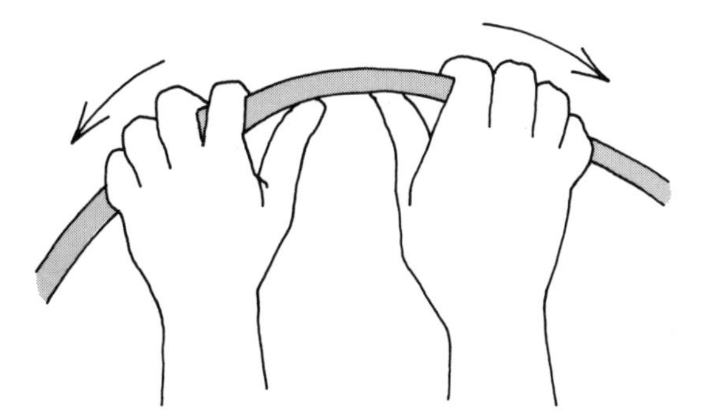

Continue *in this fashion until bend is complete. Don't be in a hurry—pipe is no good to you if it's crimped.*

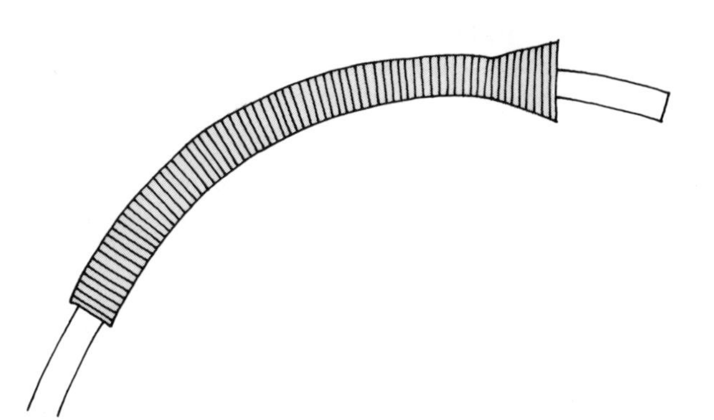

A special bending spring *fits over copper tubing and reinforces its walls to minimize chances of damage.*

Plastic pipe and tubing

Plastic pipe is lightweight, inexpensive, and easy to cut and fit, making it ideal for homeowner installation. It's also self-insulating, and resistant to weather, strong chemicals, and electrolytic corrosion, which are problems with metal pipe. Its smooth "non-wettable" interior surface provides less flow resistance, both initially and after prolonged use.

Plastic pipe comes in both rigid and flexible types. There are three types of rigid pipe: CPVC (chlorinated polyvinyl chloride) used for hot water supply systems; PVC (polyvinyl chloride) used for cold water supply and DWV systems; and ABS (5¢ if you can say it: acrylonitrile-butadiene-styrene) used for DWV systems. Flexible plastic types are polyethylene and polybutylene tubing, used for outdoor sprinkler and irrigation systems (in some cases, polybutylene tubing can be used for hot and cold water supply systems, but check first with your local building department on its use).

Plumbing codes do not universally agree on where and how plastic pipe may be used, so check local codes before installing. Plastic pipe also has various pressure ratings (or

(Continued on next page)

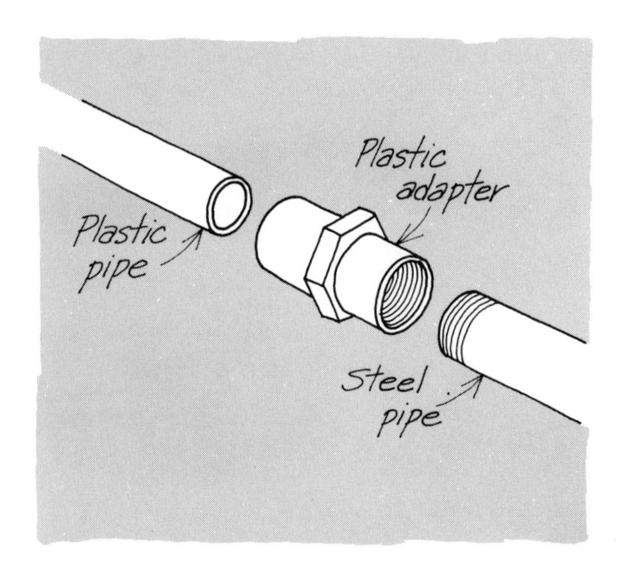

Some plastic fittings *are adapters that connect with pipe of other materials. Here threaded end fits to steel.*

schedules), so use the schedule number prescribed by your local building department.

Because pressure ratings for plastic pipe are lower than those of metal pipe, they're less likely to withstand line surges (sudden increases in water pressure) in the water supply system. Water is normally delivered at 50 psi, which poses no danger to these pipes. But full flow and rapid turn-off at fixtures (washing machines and dishwashers do this automatically) can momentarily raise the water pressure to as much as 500 psi. To eliminate line surge problems, install air chambers or shock absorbers at all appropriate fixtures (air chambers are discussed on page 27). Also, when using CPVC for hot water supply, replace the temperature-pressure valve on your water heater to fit temperature-pressure ratings for CPVC (standard is 100 psi @ 180°F.).

Cutting plastic pipe. Before you make any cuts, see page 63 for how to measure a pipe run. You can cut rigid plastic pipe with a hacksaw (24 or 32-tooth blade), a fine-toothed handsaw, or a fine-toothed power saw. With the handsaw, use a miter box to get a square cut; the power saw should have guides to insure a square cut. You can also use a pipe cutter with a special blade for plastic pipe.

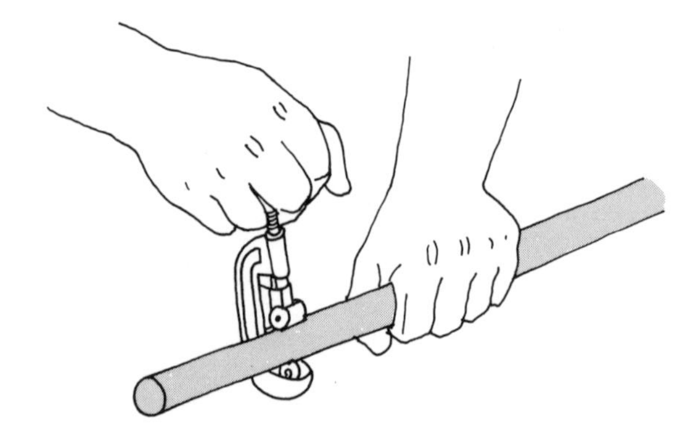

Special pipe cutter *for plastic gives good, even cut. Blade on this cutter is narrow to prevent large burrs.*

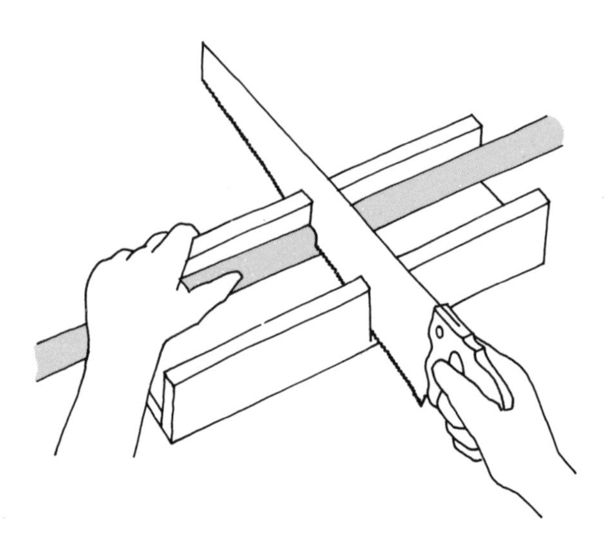

You can also cut plastic pipe *with a regular handsaw. To insure a square cut, place it in a miter box, hold securely.*

Fitting plastic pipe. Cut off all burrs, inside and out, with a knife or a reamer. Clean the end of the pipe with a rag.

Do a trial fit of the pipe and fitting. If the pipe won't fit, file or sand it down. When the pipe and fitting go snugly together (not so loose that the fitting will fall off), remove the gloss on the outside of the pipe and on the inside of the fitting, using fine sandpaper or a liquid cleaner made for the purpose.

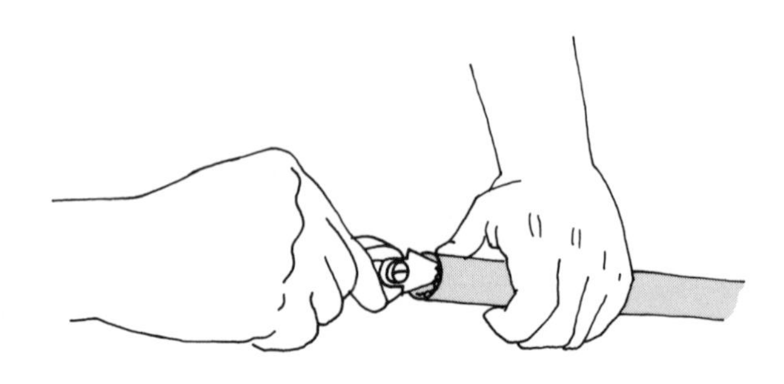

After cutting plastic pipe *you'll find burrs, just as on metal. You can remove them with a sharp knife instead of a reamer.*

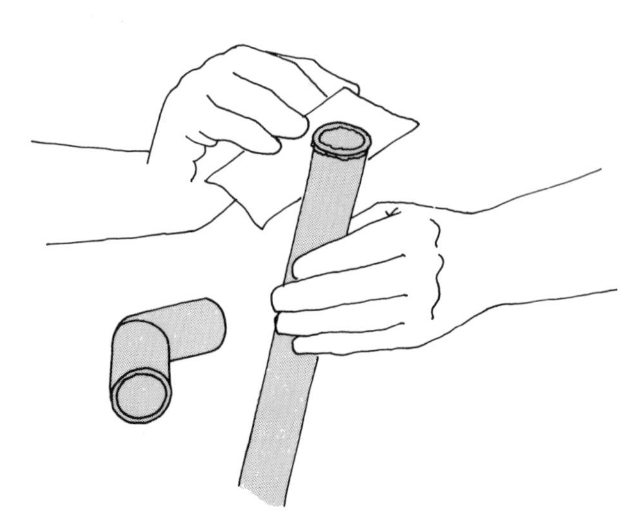

Once you've removed any burrs *from inside and outside the pipe end, clean it off very thoroughly with a dry cloth.*

For bonding the pipe and fitting, use a special cement and a non-synthetic bristle brush. Some cement comes in a container with an applicator brush. For pipe ½ inch or smaller, you can use a ½-inch brush. Use a 1-inch brush for pipe up to 2 inches. For larger pipes use a brush at least half the nominal pipe size. Be sure that you get the right type of cement for the kind of plastic you're using.

Before cementing the pipe and fitting, you should know exactly how the finished run will line up. It's a good idea to mark the pipe and the fitting beforehand and line up the two marks when cementing.

Brush a light coat of cement on the inside of the fitting; apply a heavy coat to the pipe end. Put the two pieces together immediately. After you give them a quarter turn to spread the cement evenly, line them up precisely. Wipe the excess cement from around the lip of the fitting. (The bead of excess cement should be uniform all around the fitting. If there's no bead or an uneven bead, the joint needs more cement; if there's a very heavy bead, there's probably too much cement in the fitting.) Check inside the fitting with your finger to make sure no cement is blocking it.

Hold the fitting and pipe together for about a minute. Then wait at least an hour before you put any water into the pipe. (If air temperature is between 20° and 40° F., wait at least 2 hours; between 0° and 20° F., wait at least 4 hours.)

Polyethylene tubing. Polyethylene tubing is made for outdoor use—most often for wells and sprinkler systems. It isn't suitable for hot water. It comes in three grades, rated at 125 psi, 100 psi, and 80 psi. This last one is good for most residential underground watering systems. (For information on planning and installing sprinkler systems, see page 59-61.)

You can cut polyethylene tubing with a knife or a hacksaw. Making perfectly square cuts is not crucial.

Fittings for polyethylene are ridged plastic, held in place by screw clamps. Adapters can join polyethylene pipe to other kinds of pipe (sprinkler heads are most often steel).

With polyethylene systems you can take apart and reassemble fittings as often as necessary. If you have difficulty getting a fitting to come apart, pour hot water over the end of the tubing to soften it.

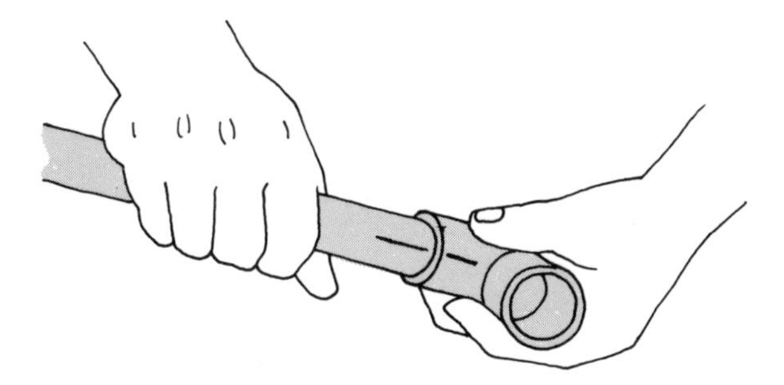

Place fitting *over the end of the pipe (it should fit snugly) and mark both pipe and fitting to show exact placement.*

Flexible polyethylene *plastic pipe is very easy to work with. You can cut it with a saw or, shown here, with a sharp knife.*

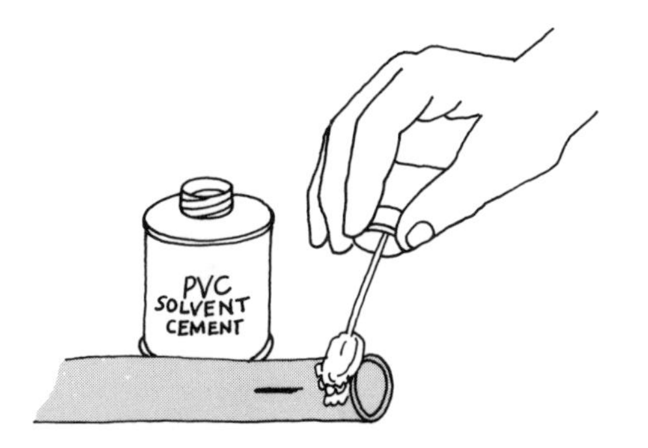

Apply a thick coat *of special plastic solvent cement to the outside of the pipe end. It dries fast, so don't dawdle.*

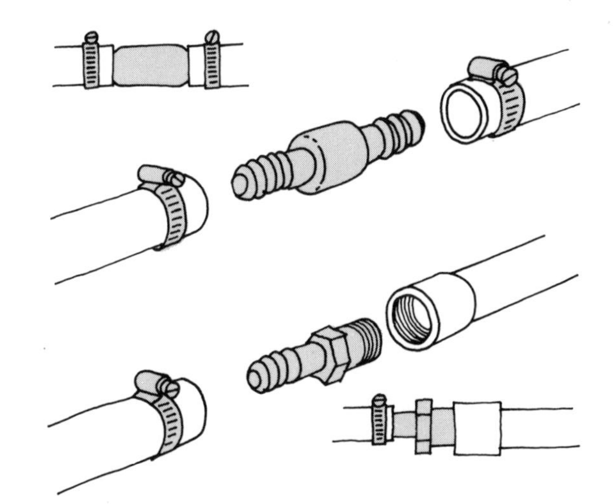

Top: *Ridged plastic fitting and clamps join two lengths of polyethylene pipe.* **Bottom:** *Adapter joins plastic, steel pipes.*

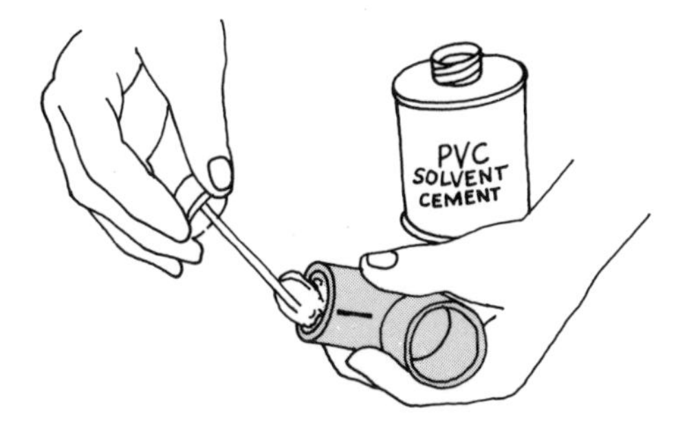

Coat inside of fitting *with very light coat of solvent. Once you begin cementing, you have less than a minute to finish.*

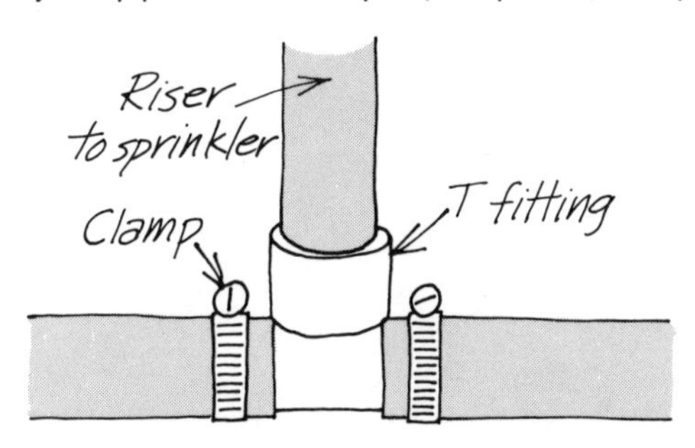

Special adapter T *joins polyethylene pipe with threaded galvanized steel pipe (called a "riser") for sprinklers.*

Residential DWV pipes may be plastic, copper, iron, or steel. DWV fittings are called sanitary fittings. They differ from standard fittings in that they have no interior shoulder that could catch waste.

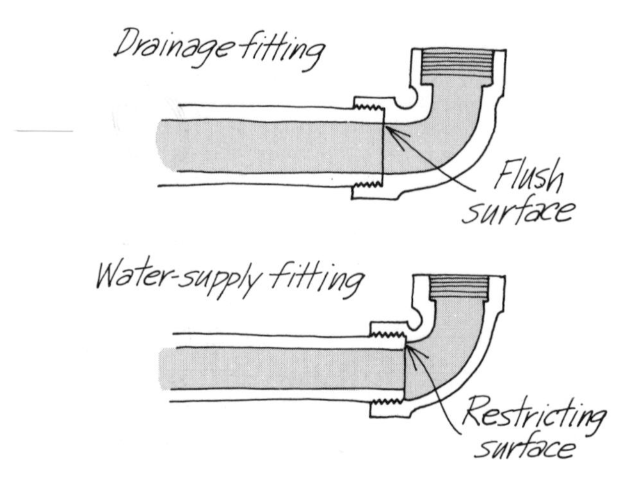

Cut-away view *of drainage and supply fittings. Drainage fitting has no shoulder to block smooth flow of waste.*

Horizontal runs of DWV pipe are installed with a slight pitch or slope. Normal pitch is about ¼ inch per foot of horizontal distance the pipe travels. When replacing DWV pipe or adding on to the system, you'll need to calculate pitch carefully.

DWV pipes are larger than supply pipes. Branch drain pipes from tubs, showers, sinks, and lavatories are normally 1½ or 2-inch pipe; branches from toilets, 3-inch pipe; soil stacks and house drains, 3 or 4-inch pipe; house sewers, 4-inch or larger.

To check for leaks in the DWV system when you've completed fitting, you'll need to rent some special plugs that will close the system at the point where the drain and the house sewer meet. Then go up onto the roof with a garden hose and fill the system with water down through a vent until water is up to the vent. Check all joints for leaks.

Plastic and copper DWV pipes

Lightweight copper and plastic are particularly useful for the large sizes required for DWV pipes. Adapters are available to fit both plastic and copper to other kinds of pipe. Copper pipe and fittings in the DWV size are very expensive. If your code permits, save some money by using thin-walled type M copper pipe. If your code permits its use, plastic DWV pipe is even more economical.

Measuring DWV. Measure a run from the face of one fitting to the face of the other. To allow for fittings, measure the distance from inside the fitting where the pipe stops to the fitting's outside edge. Add that distance to the face-to-face distance between fittings to arrive at the correct length to cut the pipe (see page 63).

Joining and cutting DWV. Both plastic and copper are cut and joined as described on pages 68-73. For soldering the larger sizes of copper pipe, you'll want a bigger torch than that used for supply-size pipe. And you'll need a bigger brush for applying the cement to plastic pipe.

Cast-iron DWV pipes

There's a good chance that the DWV pipe in your house is cast iron. Cast-iron pipe is strong, highly resistant to corrosion, and quieter than plastic pipe. Though it's less expensive than copper, its heaviness makes it quite difficult to work with.

Bell and spigot. Bell and spigot cast-iron pipe takes its name from the bell-shaped hub at one end of the pipe and the lip called a "spigot" at the other end. The spigot end of one pipe fits into the bell end of another when standard lengths are used. Otherwise, the spigot is cut off when the pipe is cut to size. The bell and spigot type must be caulked when fitted together, requiring many special tools, oakum,

Plastic DWV pipes *as they look in place under a house. Pitch is exaggerated because house is on a slope.*

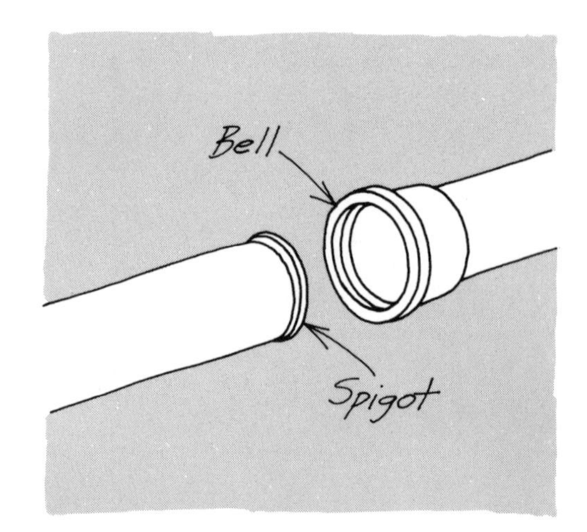

Small lip, *or spigot, on male pipe ends, left, fits into bell on female pipe end, right. Fitting needs caulking, molten lead.*

Cutting cast-iron pipe

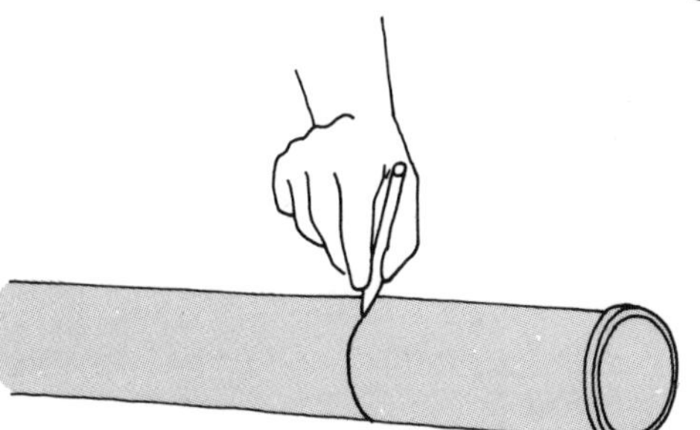

Step 1. *With chalk or crayon, mark precisely where to cut.*

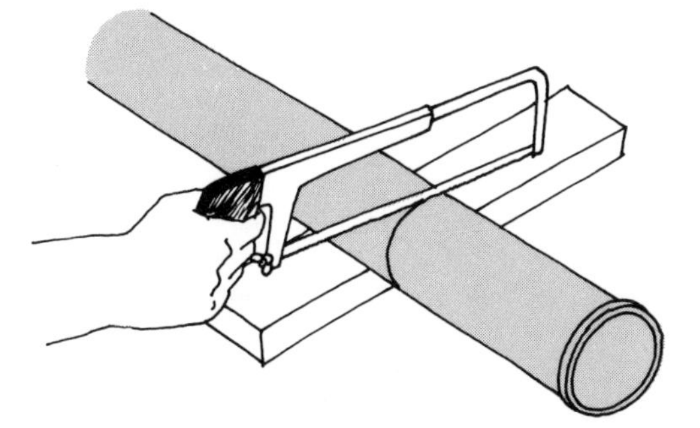

Step 2. *Score all around pipe with hacksaw to ¼-inch depth.*

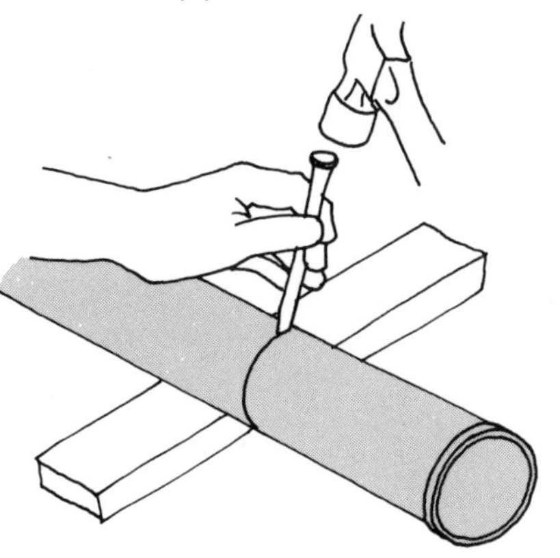

Step 3. *Make a deeper cut in supported pipe with a chisel.*

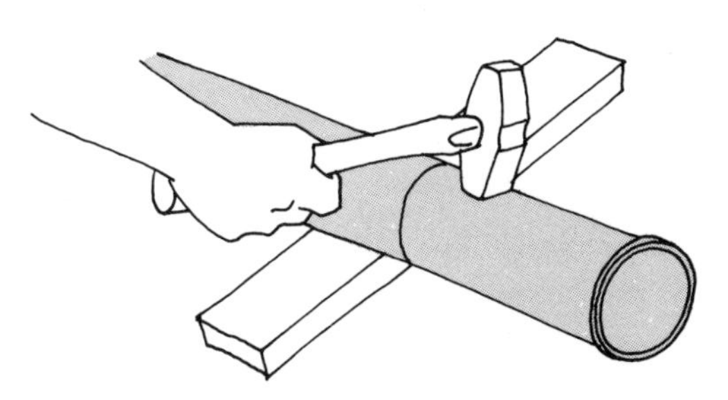

Step 4. *Finish cut by tapping all around with hammer.*

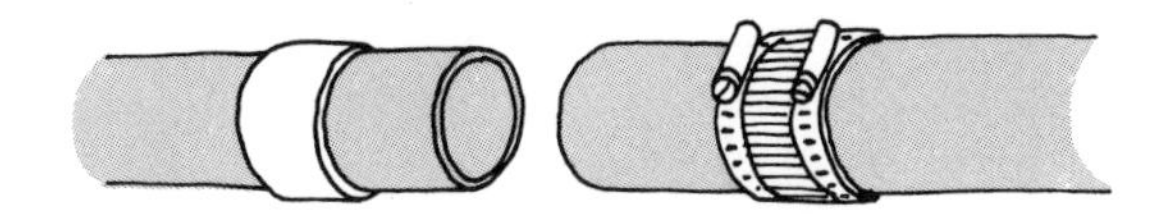

Step 5. *Fit clamp and gasket onto pipe.*

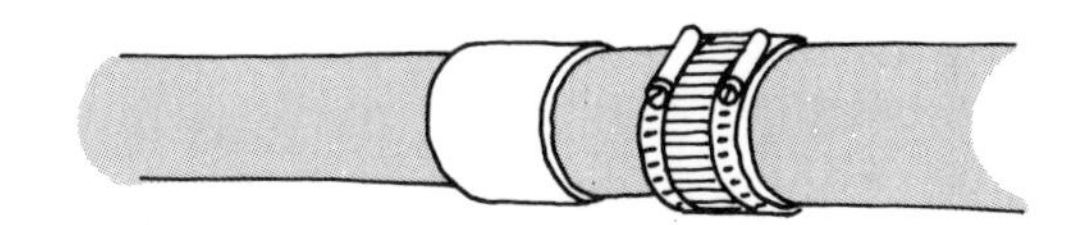

Step 6. *Fit pipe ends together.*

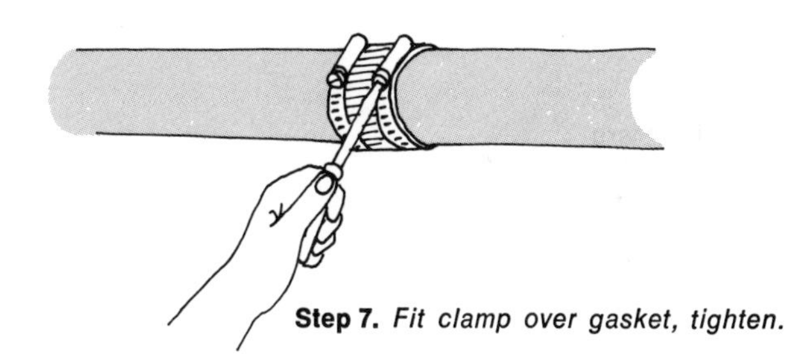

Step 7. *Fit clamp over gasket, tighten.*

and molten lead—making it a forbidding task for all but the most experienced home plumber.

Hubless cast-iron pipe. A recent development for fitting cast-iron pipe simplifies the use of this heavy material. Instead of tackling all the complications and mess of lead work, you can join this hubless pipe with neoprene rubber gaskets and screw-on steel clamps. Since the pipe ends fit flush with each other, you don't have to make any allowance for joints when measuring runs.

The neoprene gasket fits over the end of one pipe, the clamp over the end of the other. Fit the pipe ends together and center the gasket; then pull the clamp over the gasket and tighten it. You can fit hubless into an existing run of bell and spigot by cutting off the hubs when you remove the run. Adapters are available for fitting hubless cast-iron pipe to pipes made of other materials. If you're going to install cast-iron, hubless is your best choice.

Compression fittings. Another fitting for cast-iron pipe uses a lubricated gasket in the bell end. This kind of fitting, called a compression fitting, takes a spigotless pipe end in the gasket.

To install a compression fitting, first fit the lubricated rubber gasket into the hub. Then force the pipe it will be joined with into the gasket, forming a watertight seal. (Check your code before using this kind of fitting.)

The Durham system

Large, threaded, galvanized DWV pipes are used in the Durham system. Because the pipes are so heavy, cutting and threading and joining are particularly difficult for the home plumber. The easiest method, and the one that will save you much trouble and expense, is to measure everything precisely and have the pipes cut and threaded at your plumbing supply store.

TAKING APART A PIPE RUN

When you're replacing leaky pipes, fittings, or valves, the first step is to disassemble the section that includes the faulty part.

Before you begin, turn off the water and drain the system by opening the special drain valve (if you have one) or the lowest faucet in the house.

Removing galvanized pipe

If you need to take apart a run of galvanized iron or steel pipe and there's no union in the run, you'll have to saw the pipe in half. (Unscrewing the pipe at one end would mean tightening it at the other.) With a fine-toothed hacksaw (24 or 32-tooth blade), cut completely through the pipe. (Have a bucket or absorbent cloth in place under the cut to catch any spill.) Brace the pipe while you saw to prevent excess motion that would strain joints. Avoid letting the pipes sag when cutting is finished. Using two pipe wrenches—one on the pipe and one on the fitting—unscrew first one section of the cut pipe and then the other. If you have trouble unscrewing the pipes, apply liberal doses of oil to the joints.

To replace the section of pipe, you'll need two pieces of pipe and a fitting called a union. When measuring, allow for all fittings in the run. (See pages 64-65 for information on installing galvanized pipes.)

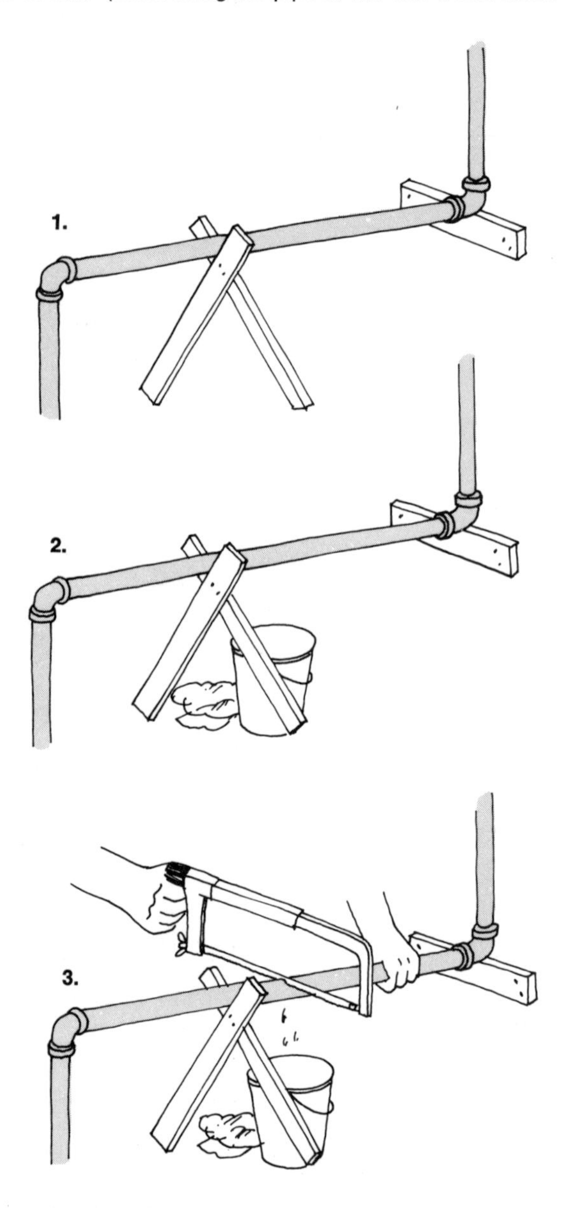

First, *devise a way to brace pipes. Sagging will cause strain that could damage other pipes, fittings. Turn off the water and then place a bucket or a pan under the spot to be cut to catch spillage. Have rags ready, too.*

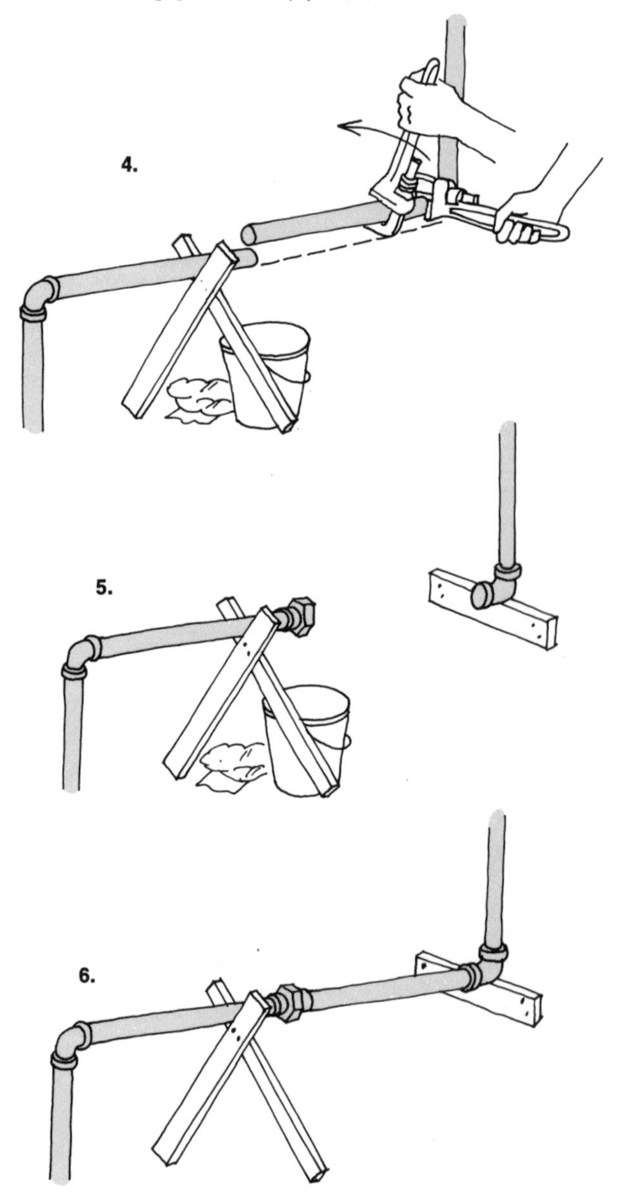

After cut is made, *use two pipe wrenches—one on the fitting, one on the pipe—turning opposite directions, to remove pipe. A union must be used to install new run. Pipes screw into fittings at either end first, then screw into union.*

Removing copper pipe

To disassemble a run of rigid copper pipe with soldered fittings, you just need to melt the solder with a torch if there is enough give in the run to pull pipe ends free of fittings. If not, you will have to cut the pipe first.

Reassemble a run of rigid copper pipe that has been cut by using a union. If you've unsoldered the joints, check the fitting at the run's other outlet to make sure the joints remain intact. Since the presence of any water at all in the pipes will hinder a successful soldering job, dry the pipes as much as possible. Then stuff the ends with plain white bread. Apply flux, reassemble, and solder. The bread will absorb any moisture and, after the job is finished, the bread will soften and wash away when the water is turned on.

You can disassemble a run of copper tubing with soldered fittings by melting the solder with a torch and pulling the pipe ends out of the fittings. The tubing will give enough to allow pipe ends to slip out of the fittings. Pull gently so you don't bend or damage the pipe. Reassemble the run by slipping the tubing into the fittings and resoldering without a union.

Compression and flare fittings are even easier to take apart—just unscrew them and pull the pipe ends out.

Removing plastic pipe

When a piece of plastic pipe is cemented into a fitting, there's no way to get the two apart. To disassemble a run, you must cut off the fittings.

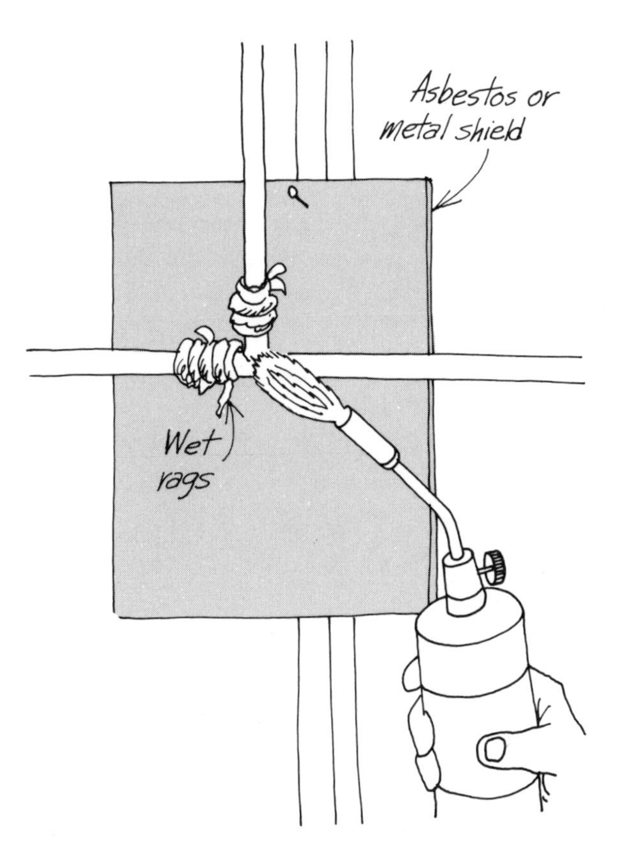

When unsoldering *copper pipe, shield wood, wrap other fitting outlets with wet rags so they don't come apart, too.*

With a fine-toothed hacksaw, cut through the pipes just outside the fittings on each end of the run to be disassembled. Because you've had to cut out the old fittings, the new run will be somewhat longer; measure carefully before reinstalling. Use a union to complete the new run.

Removing cast-iron pipe

The tool pictured makes quick work of removing a length of cast-iron pipe. It's available at equipment rental stores. If the pipe to be removed is situated so you can't make a full 360° turn around it, be sure the cutter you use has a ratchet head.

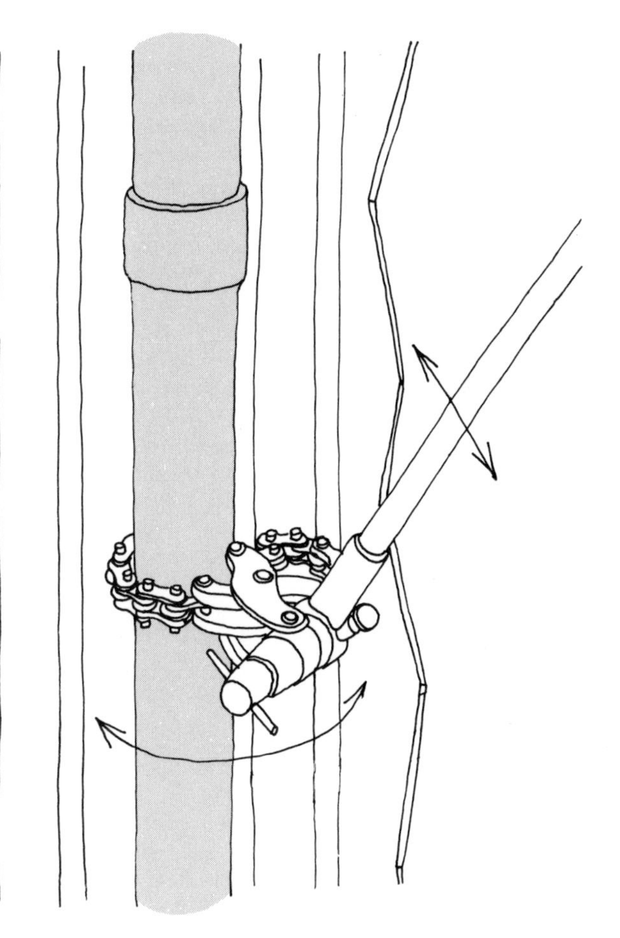

Special cutter *for cast-iron pipe. Its ratchet head makes it especially useful in tight quarters. Available at rental stores.*

You can also cut through cast-iron pipe using a hammer and cold chisel or a hacksaw, but this method takes much longer. With a piece of chalk, mark the pipe at the cutting point all around its circumference. Cut about 1/16 inch into the pipe all along the chalk line. Tap the pipe repeatedly with a hammer. If it doesn't come apart, you may have to cut all the way through. Be sure to have someone supporting the pipe while you cut so it doesn't fall on you.

If you reinstall with cast iron, use hubless. Measure exactly the old pipe's diameter and confer with your supplier when buying the new pipe—sizes of hubless and bell and spigot may not correspond exactly.

Plumbing language

ABS. Rigid plastic pipe. Acrylonitrile-butadiene-styrene.

Adapter. Fitting for pipes of two different materials.

Aerator. Sievelike device on spigot end; mixes air with water flow.

Air chamber (or air cushion or water hammer arrester). Device attached to supply pipes near outlets to prevent water hammer. (See page 27.)

Allen wrench. Hexagonal-end wrench, used to remove and install faucet valve seats.

Anti-siphon valve. Installed on supply line to prevent siphoning of contaminated water back into water-supply system.

Auger (or snake). Springlike tool forced into waste lines to break up blockages.

Back venting or reventing or secondary venting. Connecting a fixture with nearby main vent instead of venting directly out the roof.

Ballcock. Assembly inside toilet tank that connects with water supply and controls flow of water into tank.

Balloon bag. Device that attaches to garden hose to loosen blockage in clogged drain. (See page 22.)

Basin wrench. Tool designed to install or remove hard-to-reach locknuts holding deck-mounted faucets onto sink.

Bell and spigot cast-iron pipe. One end has a bell-shaped hub, the other end a lip called "spigot." Spigot end of one pipe fits into bell of another. Caulked with oakum, sealed with molten lead.

Bonnet. Casing for wall-mounted bath and shower faucets. Screws into faucet body behind the wall.

Cam. Device in tipping-valve faucets that controls water flow by tipping valve stems.

Cap. Fitting with a solid end used for closing off pipe end.

Cap nut. On compression faucet, chrome-plated nut that holds stem assembly onto faucet body.

Caulking. Material used to create a water-tight seal.

Center-to-center. In mounting faucets: distance between centers of holes on sink deck. In pipefitting: distance between centers of two consecutive fittings in a run.

Cleanout. Opening providing access to drain line or trap under sink, closed with threaded plug.

Closet auger. Tool for clearing blockages in toilet.

Closet bend. Drainpipe that joins with toilet bowl outlet at one end, drain line or soil stack at the other.

Code. All legal requirements for plumbing installation.

Compression fitting. For copper tubing. (See page 70.)

Continuous waste. Waste pipe from two or more fixtures, using same trap.

Copper pipe. Lightweight, rigid pipe joined by soldered or threaded joints. Weight and ease of fitting make it good choice for home plumber.

Copper tubing. Flexible tubing joined by soldered, flare, or compression fittings.

CPVC. Plastic pipe. Chlorinated polyvinyl chloride.

Drain cock. Simple valve connection at lowest part of water-supply system that can be opened to drain the system.

Drum trap. Trap occasionally used for tubs and showers instead of curved pipe sections; a cylindrical drum with inlet and outlet at different levels.

Durham system. Drain-waste-vent system using threaded galvanized pipe.

DWV (Drain-waste-vent). System that carries away waste water and solid waste, allows sewer gases to escape, maintains atmospheric pressure in drainpipes.

Elbow. Fitting used for making turns in pipe runs (for example, 90° elbow makes right-angle turn). A street elbow has one male end and one female end.

Escutcheon. Decorative piece that fits over faucet body or pipe coming out of wall.

Female. Pipes, valves, or fittings with internal threads.

Filler tube. On ballcock assembly, tube through which water enters toilet tank.

Fitting. Device used to join sections of pipe.

Flange. Flat fitting with holes to permit bolting together (toilet bowl is bolted to floor flange).

Flare fitting. Female fitting used on copper tubing. (See page 70.)

Flashing. Device that fits over vent pipe on roof to prevent water from entering house through vent opening.

Float arm. Wire arm that connects float ball at one end to ballcock assembly at other in toilet tank.

Float ball. Large copper or plastic ball that floats on surface of water in toilet tank, descends and ascends with water level.

Float valve. Valve in ballcock assembly that controls flow of fresh water into toilet tank.

Flue. Large pipe through which fumes escape from gas-fired water heater.

Fluorocarbon tape. Special tape used as a joint sealer in place of pipe joint compound.

Flush ball. Device that fits over outlet at bottom of toilet tank. When raised, permits water in tank to flow downward into bowl.

Flush valve. Controls flow of water from tank into bowl (comprises flush ball and valve seat at outlet).

Flux. Paste applied to copper pipe, tubing before soldering. Prevents oxidation when heat is applied to metal.

Galvanized pipe. Iron or steel pipe coated with zinc to retard corrosion. Fittings are threaded.

Gasket. Device (usually rubber) used to make the joint between two valve parts watertight.

Gate valve. Valve with tapered member at end of stem; acts as a gate to control flow of water.

Globe valve. Valve with washer at end of stem that fits into valve seat to stop flow of water.

Graphite packing. Wirelike material to be wrapped around faucet stem to prevent leaking.

Hose bib (or hose cock). Valve with male threaded outlet for accepting hose fitting.

Hubless cast-iron pipe. Cast-iron pipe that is joined using rubber gaskets and clamps, making it much easier to work with than bell and spigot, which uses molten lead.

Increaser. Fitting having larger opening at one end for accepting larger diameter pipe.

J-bend. J-shaped piece of drainpipe used in traps.

Joint. Point at which two sections of pipe are fitted together.

Locknut. Nut fixed onto one piece (for instance, flexible connector for water heater) and screwed onto another piece to join the two.

Male. Pipes, fittings, and valves with external threads.

Nipple. Short length of pipe (under 12 inches) with male threads on both ends for joining fittings.

Oakum. Stranded hemp used in making bell and spigot joints watertight.

O-ring. Narrow rubber ring used in some faucets instead of packing to prevent leaking around stem; also used with swivel-spout faucets to prevent leaking at base of spout.

Packing. See "Graphite packing."

Packing nut. Nut screwed down onto faucet stem, holding packing tight.

Penetrating oil. Used to help loosen threaded joint in which corrosion has fused fittings.

Phillips screwdriver. For cross-shaped slot in Phillips screw.

Pipe cutter. Various tools designed specifically for making perfectly square cuts on pipe. Cutters for copper, plastic, galvanized, cast iron are available.

Pipe joint compound. Sealing compound used on threaded fittings (apply to male threads).

Plug. Closed-end, male-threaded fitting for closing off pipe end that has female threads.

Polyethylene tubing. Flexible plastic pipe; most often used for underground sprinkling systems.

Power auger. Auger used for clearing blockages in house sewer. Available at rental stores.

Pressure-reducing valve (temperature/pressure-reducing valve). Safety valve for water heater; lets water and steam escape.

P-trap. Fixture trap in the shape of the letter P.

PVC. Rigid plastic pipe. Polyvinyl chloride.

Reamer. Tool that fits into pipe ends and is used to grind off internal burrs caused by cutting pipe.

Reducer. Fitting that connects pipe of one diameter with pipe of smaller diameter.

Reventing. See "Back venting."

Rim. Stainless steel device that fits around outside edge of some kinds of sink, holding sink onto countertop.

Riser. Vertical run of pipe.

Saddle tee (or T). Fitting for copper or galvanized pipe that is bolted onto pipe, eliminating cutting and threading or soldering. (See page 55.)

Sanitary fittings. Fittings that have no inside shoulders to block flow of waste; used to join DWV pipe.

Seat. Valve part into which washer or other piece fits, stopping flow of water.

Secondary venting. See "Back venting."

Sewer rod. Metal rod forced into sewer to break up blockages.

Sewer tape. Flexible metal tape forced into sewer to break up blockages.

Siphoning. When vacuum occurs in a section of pipe, nearby water is pulled into it. Toilet functions on a siphoning principle. Siphoning is not desirable other places, though. It could cause draining of fixture traps. Venting acts to prevent siphoning.

Slipnuts. Nuts that are not fixed but can move up and down the pipes, allowing for adjusting length for proper connection.

Sludge. Solid waste matter that settles to the bottom of a septic tank.

Soil stack. Large DWV pipe that connects toilet to house drain and also extends up and out the house roof, its upper portion serving as a vent.

Solder. Soft metal wire (tin and lead) used as a bonding agent to join copper pipe.

Solvent cement. Compound used to join rigid plastic pipe.

Spud wrench. Special tool for working with particularly large nuts, such as those that hold sink strainers in place or those at outlet pipes of wall-hung, two-piece toilets.

Standpipe. Special drainpipe for automatic washer.

Stop-cock. Underground valve near property line used for shutting off water in emergencies.

Strap wrench. Tool used for working with pipe whose finish is easily marred. Uses strap instead of metal jaws.

Stub-outs. Portions of supply pipe that stick out of wall.

Sweating. 1) Another name for soldering joints. 2) Accumulation of moisture on pipes, tanks, caused by condensation when cooler surface of pipe or tank meets warm air.

Swivel-head washer. Washer that has a swivel base fixed to a clip top.

Tank ball. See "Flush ball."

Tee (or T). A fitting with three openings, shaped like a T.

Temperature/pressure-reducing valve. See "Pressure-reducing valve."

Tempering valve. Valve that mixes a small amount of hot water with cold water entering toilet tank in order to prevent sweating tank.

Threading die. Tool used for cutting threads into pipe.

Toggle bolt. Bolt with two hinged wings; can be used to repair leaks in water tanks.

Trap. Device (most often a curved section of pipe) that holds a water seal to prevent sewer gases from escaping into home through fixture drain.

Trap arm. Section of trap that connects J-bend with drainpipe behind the wall.

Union. Fitting that joins two lengths of pipe and permits assembly, disassembly without taking entire section apart.

Valve. Device that controls flow of water.

Valve seat. See "Seat."

Valve seat dresser. Tool used to grind down burrs on valve seat.

Valve seat wrench. Hexagonal-end wrench inserted into hexagonal opening in valve seat for installing or removing seat.

Water hammer. The sound of pipes shuddering and banging, resulting from lack of air cushion.

Y. Fitting with three outlets in the shape of the letter Y.

Index

Photographers

Paul Aller: 62 top right. **Roger Flanagan:** 62 bottom left, right. **Ells Marugg:** 4 left, right; 10; 19; 20; 21; 22; 23; 27; 43; 68; 69; 74. **Norm Plate:** 62 top left. **Darrow Watt:** 4 center.